Steck-Vaughn

Strength in Numbers™

Algebra

Level 8

www.steck-vaughn.com

Acknowledgments

Editorial Director: Diane Schnell
Supervising Editor: Donna Montgomery
Associate Director of Design: Cynthia Ellis
Design Manager: Deborah Diver
Production Manager: Mychael Ferris-Pacheco
Production Coordinator: Susan Fogarasi
Editorial Development: Words and Numbers
Production Services: Mazer Corporation

Photo Credits: Cover (fireworks) ©Bill Ross/CORBIS; cover (inside computer) ©Hugh Burden/Getty Images; cover (sunflowers) ©Alvis Upitis/Getty Images; p.5 ©Lois Ellen Frank/CORBIS; p.10 ©Jeffrey Sylvester/Getty Images; p.35 ©Bill Ross/CORBIS; p.67 ©Hugh Burden/Getty Images.

Additional Photography by Corbis Royalty Free, Getty Images Royalty Free, PhotoSpin Royalty Free, and the Steck-Vaughn Collection.

ISBN: 0-7398-6251-0

1 2 3 4 5 6 7 8 9 10 WC 09 08 07 06 05 04 03 02

Contents

Unit 1: Integers and Monomials **5**

Addition of Integers with the Same Signs 6
Addition of Integers with Different Signs 8
Subtraction of Integers Using Models 10
Subtraction of Integers 12
Multiplication and Division of Integers 14
Addition and Subtraction of Monomials 16
Exponents 18
Scientific Notation 20
Multiplication of Monomials 22
Division of Monomials 24
Simplified Expressions 26
Negative Exponents 28
Rational Expressions 30
Strength Builder 32
Unit 1 Review 34

Unit 2: Properties and Equations **35**

Properties of Addition 36
Properties of Multiplication 38
Order of Operations 40
Distributive Property 42
Addition and Subtraction Equations 44
Multiplication and Division Equations 46
Multi-Step Equations 48
Equations with Fractions 50
Equations with Like Terms 52
Equations with Variables on Both Sides 54
More Equations with Variables on Both Sides 56
Addition and Subtraction Inequalities 58
Multiplication and Division Inequalities 60
Multi-Step Inequalities 62
Strength Builder 64
Unit 2 Review 66

Contents

Unit 3: Graphs and Functions **67**

Linear Equations 68
Linear Equations and Graphs 70
Slope of Lines 72
Slopes and Intercepts 74
Direct Variation 76
Tables of Solution Sets 78
Solution Sets and Graphs 80
Graphs of Inequalities with Two Variables 82
Systems of Equations 84
Linear Functions 86
Data and Functions 88
Functions and Graphs 90
Strength Builder 92
Unit 3 Review 94

Unit 4: Polynomials **95**

Polynomials 96
Addition of Polynomials 98
Subtraction of Polynomials 100
Multiplication of Polynomials by Monomials 102
Division of Polynomials by Monomials 104
Monomials and Powers 106
Binomials and Multiplication 108
Square of the Sum of Two Terms 110
Square of the Difference of Two Terms 112
Perfect Square Trinomials 114
Differences of Squares 116
Solutions to Equations 118
Strength Builder 120
Unit 4 Review 122

Cumulative Review **123**
Glossary **126**

Unit 1

Integers and Monomials

Have you ever heard

someone say, "You can't compare apples and oranges?" What do you suppose that means?

Oranges and apples are both fruits. But they are very different looking and tasting. Algebra sometimes involves comparing and combining things that are alike. Look at the following math terms. Which terms can be added? Why can't you add all the terms?

$2y \quad 5 \quad 2z \quad 4y \quad 4y^2 \quad 7y \quad 3$

Lesson 1 Addition of Integers with the Same Signs

Integers are the set of positive and negative whole numbers and zero. The **absolute value** of an integer represents the distance on the number line between zero and the integer.

Model 1

Cara's score in a game is $^{-}5$. What is the absolute value of $^{-}5$?

$^{-}10$ $^{-}9$ $^{-}8$ $^{-}7$ $^{-}6$ $^{-}5$ $^{-}4$ $^{-}3$ $^{-}2$ $^{-}1$ 0 1 2 3 4 5 6 7 8 9 10

The absolute value of $^{-}5$ = __________.

The absolute value of $^{+}5$ = __________.

A number line can help you add integers.

Model 2

After several turns, Aaron's score in a game is $^{-}3$. If he misses a question and loses a point, Aaron's new score will be the sum of his old score and $^{-}1$. **Draw this situation on the number line.**

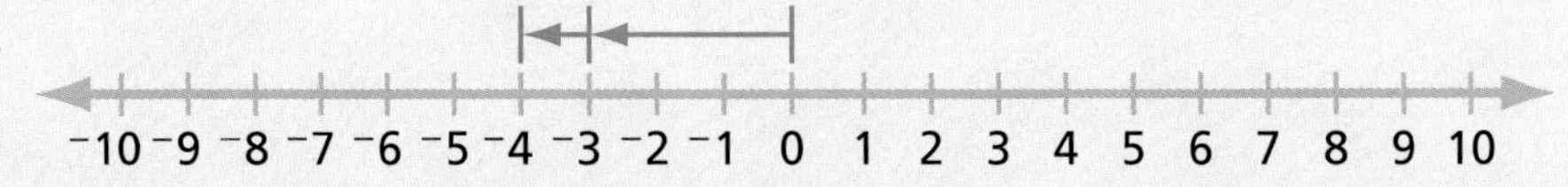

$^{-}10$ $^{-}9$ $^{-}8$ $^{-}7$ $^{-}6$ $^{-}5$ $^{-}4$ $^{-}3$ $^{-}2$ $^{-}1$ 0 1 2 3 4 5 6 7 8 9 10

Aaron's new score = $^{-}3$ + __________ = __________.

You can use what you know about absolute value to add integers with the same signs.

Model 3

Terry's scores after two rounds of a game are $^{-}6$ and $^{-}7$. What is his total score?

Find the absolute value of each number. $|^{-}6|$ = __________

$|^{-}7|$ = __________

Add the absolute values and apply the sign of both numbers to the sum.

$6 + 7$ = __________ $^{-}6 + (^{-}7)$ = __________

Write a set of rules for adding integers with the same signs.

____

____

Practice

Find the absolute value of each number.

1. $|{}^{-}4| = 4$	**2.** $|10| =$	**3.** $|{}^{-}1| =$
4. $|0| =$	**5.** $|{}^{-}20| =$	**6.** $|14| =$

Draw each equation on the number line and solve.

7. ${}^{-}1 + ({}^{-}4) = {}^{-}5$

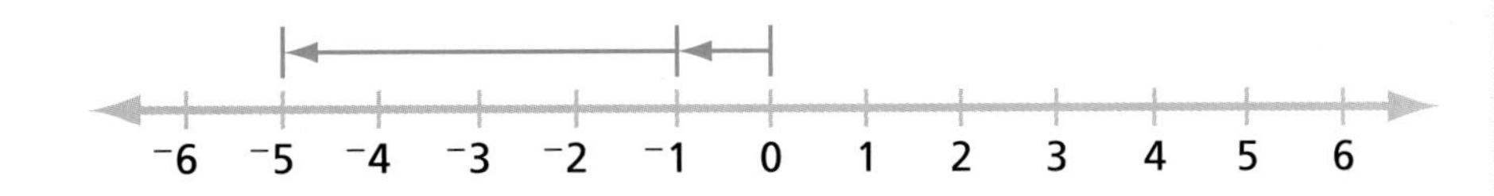

8. ${}^{-}4 + ({}^{-}1) =$

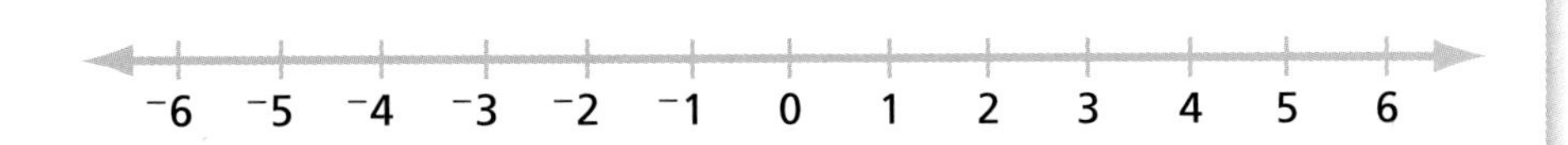

9. $3 + (2) =$

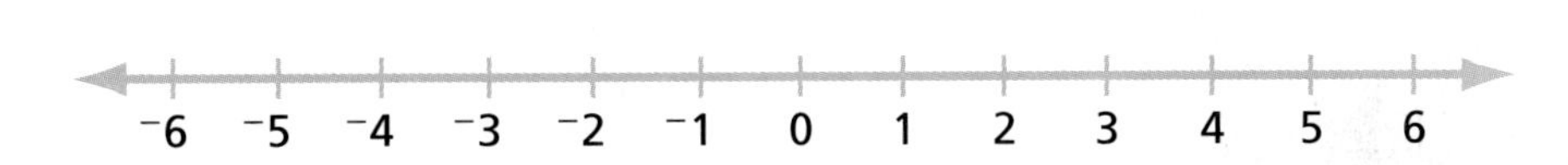

10. ${}^{-}3 + ({}^{-}3) =$

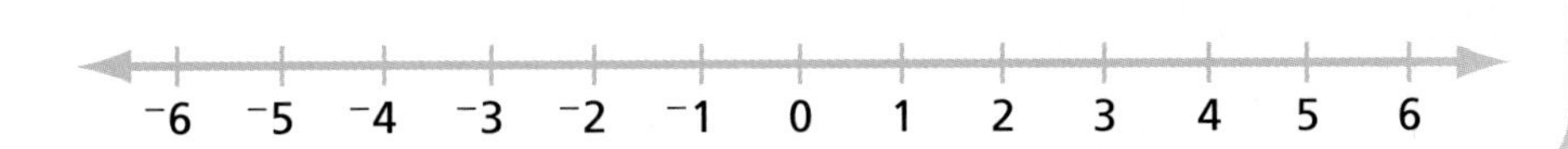

Find the sum.

11. ${}^{-}16 + ({}^{-}6) = {}^{-}22$	**12.** $3 + 21 =$	**13.** ${}^{-}17 + ({}^{-}18) =$
14. $0 + ({}^{-}7) =$	**15.** ${}^{-}15 + ({}^{-}3) =$	**16.** $42 + 0 =$
17. $26 + (5) =$	**18.** ${}^{-}11 + ({}^{-}6) =$	**19.** ${}^{-}4 + ({}^{-}28) =$

Model adding ${}^{-}2 + ({}^{-}5) + ({}^{-}1)$ on the number line below.

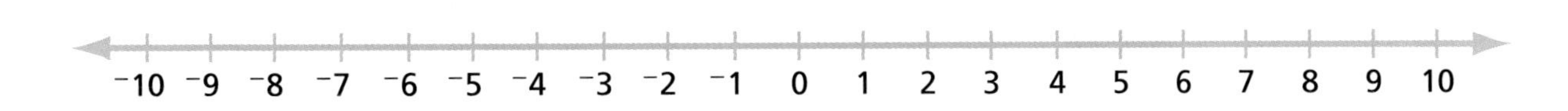

Lesson 2 Addition of Integers with Different Signs

Use a number line to model adding positive and negative integers.

Model 1

The problem $7 + (^{-}3)$ is shown on the number line below.

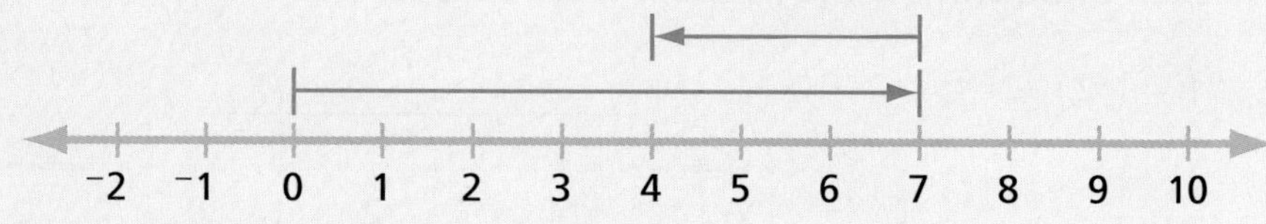

When adding a positive number, go to the right. When adding a negative number go to the left. $7 + (^{-}3) =$ ___________

Model 2

Reggie works at a state park. He starts at the rangers' station and hikes 6 miles up a trail. Then he turns around and hikes 2 miles back on the same trail. How many miles away from the ranger station is Reggie?

Model $6 + (^{-}2)$ on the number line.

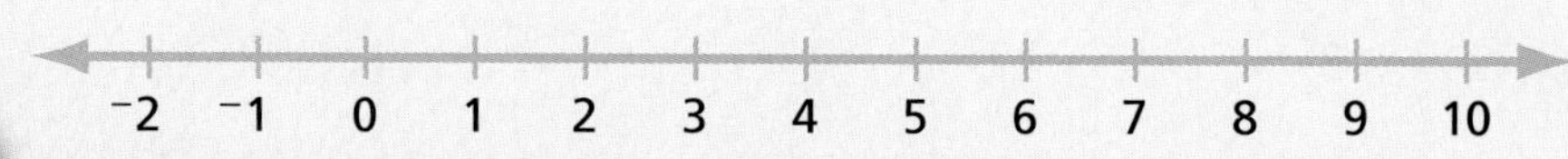

Reggie is ___________ miles from the station.

To add integers with different signs, subtract the absolute values. Then use the sign of the number with the greater absolute value.

Model 3

When Reggie left the ranger station, the temperature was $^{-}5$°F. When he returned the temperature had risen 2°F. What was the temperature when he returned?

$|^{-}5| =$ ___________ and $|2| =$ ___________

Subtract the absolute values: $5 - 2 =$ ___________

Since $^{-}5$ has the greater absolute value, the sign is negative.

$^{-}5$°F + 2°F = ___________

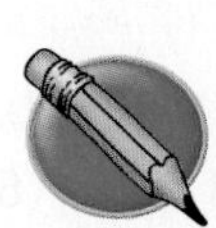

Using absolute values, explain why the sum of $^{-}10$ and $^{+}7$ is a negative number.

____

Practice

Model each equation on the number line and solve.

1. $5 + (^-9) = {}^-4$

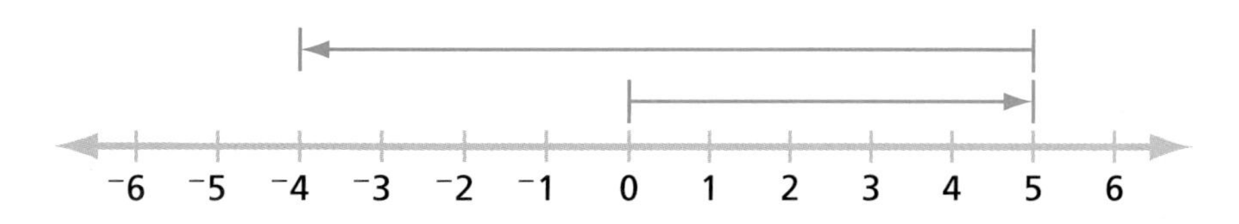

2. $^-6 + (5) =$

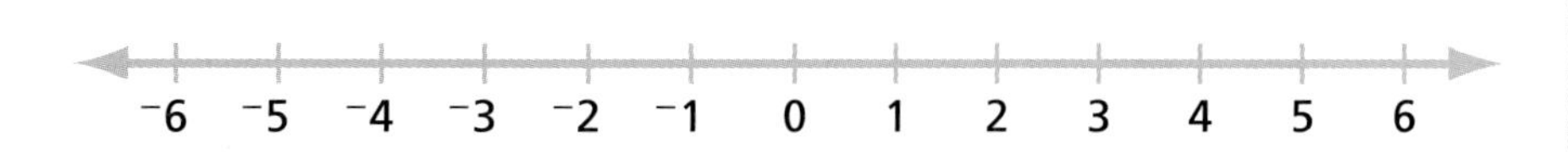

3. $5 + (^-1) =$

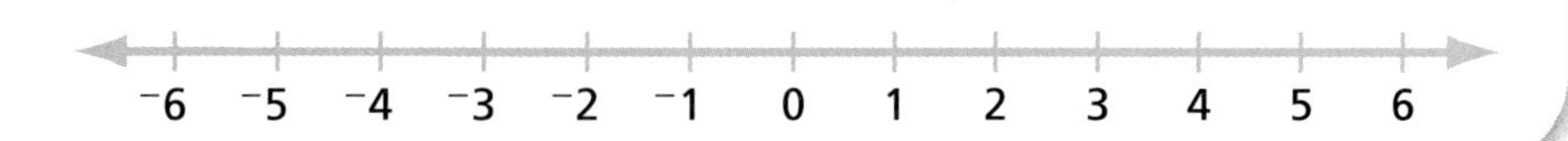

Find the sum.

4. $^-17 + 7 = {}^-10$	5. $18 + (^-20) =$	6. $^-13 + 4 =$
7. $23 + (^-4) =$	8. $0 + (^-18) =$	9. $32 + (^-13) =$
10. $^-16 + 11 =$	11. $52 + (^-45) =$	12. $70 + (^-35) =$

Write a word problem that requires the addition of positive and negative integers. Then solve the problem.

____

____

____

____

____

Lesson 3

Subtraction of Integers Using Models

Use a number line to model subtracting integers.

To subtract a positive integer, move to the **left** on the line.

To subtract a negative integer, move to the **right** on the line.

Model 1

Use a number line to model 2 − 5.

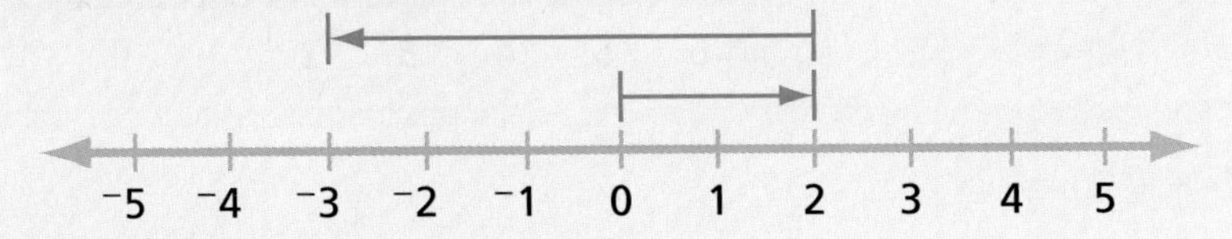

From 2, move __________ to the ____________________ to show subtraction of 5. 2 − 5 = __________

Model 2

Use a number line to model 4 − (−1).

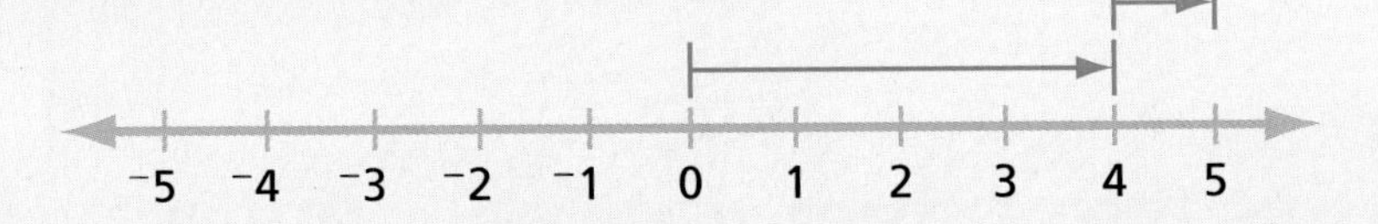

From 4, move __________ to the ____________________ to show subtraction of −1. 4 − (−1) = __________

Model 3

Use a number line to model −3 − 2.

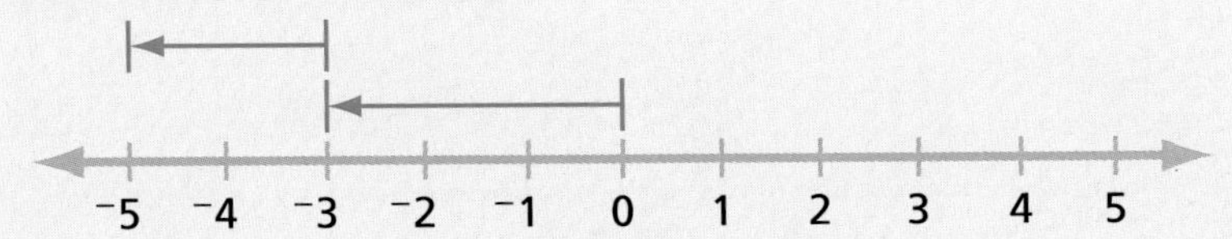

From −3, move __________ to the ____________________ to show subtraction of 2. −3 − 2 = __________

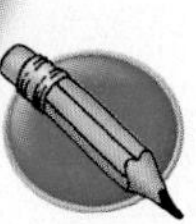

Explain how to determine from a number line if the answer to a subtraction problem is negative.

____

____

____

Practice

Model each equation on a number line and solve.

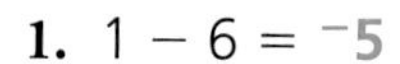

1. $1 - 6 = {}^{-}5$

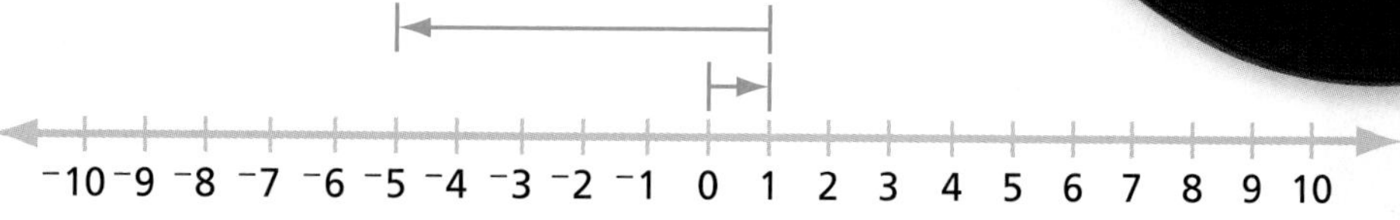

2. $2 - ({}^{-}3) =$

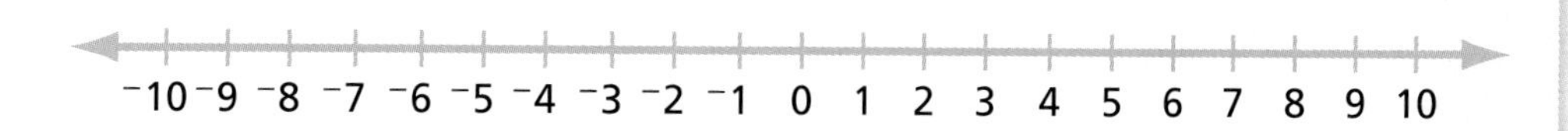

3. $7 - ({}^{-}3) =$

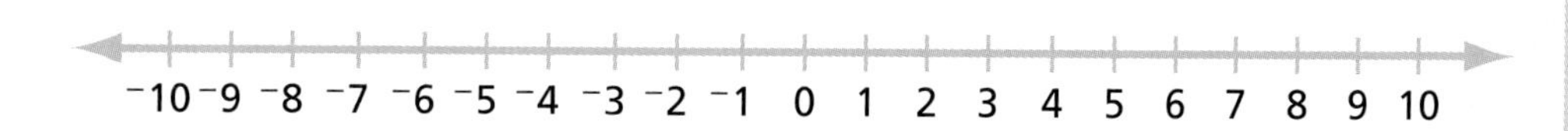

4. ${}^{-}5 - 2 =$

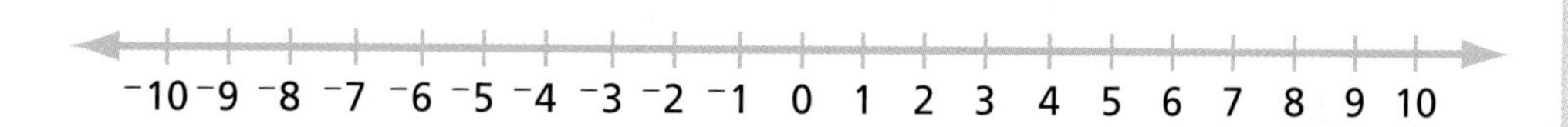

5. ${}^{-}4 - ({}^{-}2) =$

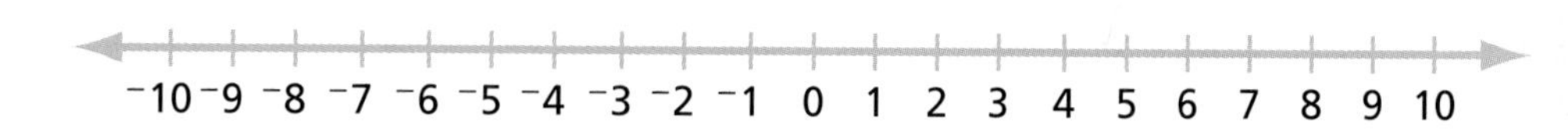

6. ${}^{-}7 - ({}^{-}3) =$

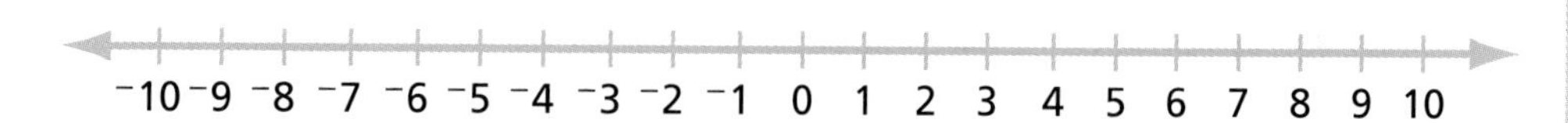

7. $5 - 5 =$

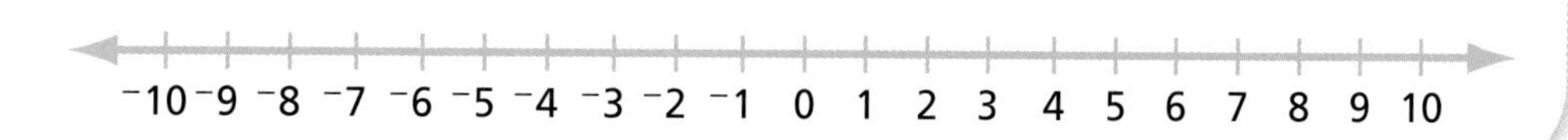

Use a number line to show why ${}^{-}1 - 2$ is less than ${}^{-}1 - ({}^{-}2)$.

Subtraction of Integers

Integer addition and subtraction are related. Subtracting an integer is the same as adding its **opposite**, or **additive inverse**. The sum of additive inverses is zero.

Model 1 **Complete the following statements about additive inverses.**

$7 + (^-7) =$ ________

$15 +$ ________ $= 0$

________ $+ (^-10) = 0$

$n +$ ________ $= 0$

Subtracting an integer is the same as adding its opposite, or additive inverse.
Subtract 9 − 4.

Model 2 Rewrite the problem as an addition problem by taking the opposite of the second number. Then use the rule for adding integers with different signs.

$9 - 4 =$

$9 +$ ________ $=$ ________

Subtract by rewriting subtraction as the addition of the opposite of the second number.

Model 3

$8 - (^-1) =$	$^-5 - 3 =$	$a - b =$
$8 +$ ________ $=$	$^-5 +$ ________ $=$	$a +$ ________
________	________	

Write one problem with subtraction of integers and explain how to rewrite the problem as an addition problem. Then solve the problem.

____

____

____

Practice

Complete the following statements.

1. $12 + \underline{^{-}12} = 0$
2. $\underline{} + 36 = 0$
3. $23 + (^{-}23) = \underline{}$
4. $0 = \underline{} + (^{-}4)$
5. $25 + \underline{} = 0$
6. $0 = {}^{-}43 + \underline{}$

Rewrite each subtraction problem as an addition problem and solve.

7. $42 - (^{-}4) = 42 + 4 = 46$
8. ${}^{-}12 - 7 =$
9. $33 - (^{-}5) =$
10. $9 - 20 =$
11. ${}^{-}2 - 16 =$
12. $0 - 29 =$
13. ${}^{-}12 - 7 =$
14. $81 - 77 =$

If the expressions are equal write = in the blank. If the expressions are not equal, write ≠ in the blank.

15. $76 - (^{-}36)$ __________ $76 + 36$
16. ${}^{-}9 - (^{-}27)$ __________ ${}^{-}9 + {}^{-}27$
17. ${}^{-}1 - 15$ __________ $1 + (^{-}15)$
18. $0 - (^{-}8)$ __________ $0 - 8$
19. $16 - (^{-}5)$ __________ $16 + 5$
20. $67 - 0$ __________ $67 + 0$

A student wrote the statement $14 - (^{-}7) = {}^{-}14 + (^{-}7)$. Explain the errors in the statement and write the statement correctly.

____

____

Multiplication and Division of Integers

You can think of multiplication as repeated addition.

Examples: $4 \times 5 = 5 + 5 + 5 + 5 = 20$ $3 \cdot 4 = 4 + 4 + 4 = 12$

Multiplication symbols can be ×, ·, or ( ). In multiplication, when the signs of two factors are alike, their product is positive.

Model 1

Is $^{-}7 \cdot {}^{-}6$ positive or negative?

The signs are the same. $7 \cdot 6 = 42$ $^{-}7 \cdot {}^{-}6 =$ __________

The product of a negative integer and a negative integer is a **positive** integer. In multiplication, when the signs are alike, the answer is

____________________.

Use repeated addition to model multiplying a positive integer and a negative integer.

Model 2

What is $4 \cdot {}^{-}5$?

$4 \cdot {}^{-}5 = {}^{-}5 +$ __________ + __________ + __________

$4 \cdot {}^{-}5 =$ __________

The product of a positive integer and a negative integer is a **negative** integer. When the signs are not alike, the answer is

____________________.

The rules for positive and negative signs also apply to division.

Model 3

Complete the following using the rules for like and unlike signs in division.

$15 \div 3 =$ __________ $15 \div (^{-}3) =$ __________

$^{-}15 \div 3 =$ __________ $^{-}15 \div (^{-}3) =$ __________

Write a pair of related multiplication and division equations to show how the rules for signs apply to multiplication and division of integers.

____

Practice

Find the product.

1. $2 \cdot (^-11) =$

$^-22$

2. $^-7 \cdot 9 =$

3. $12 \cdot 8 =$

4. $^-5 \cdot (^-5) =$

5. $13 \cdot (^-1) =$

6. $^-8 \cdot 6 =$

7. $^-2 \cdot (^-14) =$

8. $^-6 \cdot 0 =$

9. $3 \cdot (^-13) =$

10. $^-1 \cdot 15 =$

11. $10 \cdot (^-10) =$

12. $0 \cdot 9 =$

Find the quotient.

13. $16 \div (^-4) =$

$^-4$

14. $^-9 \div (^-9) =$

15. $^-10 \div 5 =$

16. $^-36 \div (^-4) =$

17. $100 \div 25 =$

18. $38 \div (^-2) =$

19. $49 \div (^-7) =$

20. $^-12 \div (^-6) =$

21. $30 \div (^-30) =$

22. $0 \div (^-5) =$

23. $^-80 \div 10 =$

24. $27 \div (^-3) =$

Is the product of three negative numbers positive or negative? Explain your reasoning and give an example.

____

____

____

____

Addition and Subtraction of Monomials

Each of the following is a **monomial**. A monomial is a single term that is a constant, a variable, or the product of a constant and one or more variables.

n $4x$ 14 y^2 $5n^3$ $^-6$ ^-8bd

Monomials like $4x$ and $5n^3$ have a coefficient and a variable. The variable represents an unknown quantity, and the coefficient states how many of the variables there are.

Examples:

$2x$	$^-4n^2$	y
coefficient / *variable*	*coefficient* / *variable*	*coefficient (1)* / *variable*

When monomials have the same variables, each to the same power, they are called **like terms**.

Combine the coefficients of like terms to add or subtract.

Model 1

Add $5n + 2n$.

$5n + 2n =$

$(5 + 2)n =$ ________

Add $3a^2 + a^2$.

$3a^2 + a^2 =$

$(3 + 1)a^2 =$ ________

When adding like monomials, add the ____________.

Model 2

Subtract $11y - 9y$.

$11y - 9y =$

$(11 - 9)y =$ ________

Subtract $8n^3 - 12n^3$.

$8n^3 - 12n^3 =$

$(8 - 12)n^3 =$ ________

When subtracting like monomials, subtract the ____________.

Model 3

When the variables in monomials are different, it is *not* possible to add or subtract them. Expressions such as $12x - 9y$ and $5a^2 + 2a^3$ are already in **simplest form** and cannot be added or subtracted.

Explain how to determine when two monomials are like terms.

____

____

Practice

Add.

1. $4k + 10k$

   14*k*

2. $10n + n$

3. $^{-}2y^2 + 7y^2$

4. $7ab + 4ab$

5. $^{-}3x + 6x + 2x$

6. $^{-}3q^3 + 5q^3$

7. $6b + (^{-}2b)$

8. $12a^2b + 3a^2b$

9. $14x + x + 2x$

Subtract.

10. $22ab - 15ab$

    7*ab*

11. $3y - 2y$

12. $17v^2 - 3v^2$

13. $5n^2 - n^2$

14. $7n - 13n$

15. $5yz^2 - 3yz^2$

16. $8w - 8w$

17. $5ac - (^{-}2ac)$

18. $^{-}16y - (^{-}2y)$

Add or subtract. If the monomials cannot be combined, write *simplest form.*

19. $5t + 9t + (^{-}3t)$

    11*t*

20. $4xy + 4x^2y^2$

21. $5n + (9n - 2n)$

22. $24c - 25c$

23. $3x + 3x - 5x$

24. $^{-}4b + 5b^3$

Explain why $5n + (^{-}n^2) + 2n^2 + (^{-}3p^2) + 7p^2 = 5n + n^2 + 4p^2$.

Exponents

Exponents provide a way to indicate repeated multiplication of a factor called the **base**. A positive integer exponent tells the number of times to multiply the base by itself.

Write each product with a common base and a single exponent.

Model 1

What is the product of $2^5 \cdot 2$?

$2^5 = 2 \cdot 2 \cdot 2 \cdot$ ______ $\cdot$ ______

$2 = 2^1 =$ ______

$2^5 \cdot 2 = 2 \cdot 2 \cdot 2 \cdot$ ______ $\cdot$ ______ $\cdot$ ______ $=$ ______

What is the product of $n^3 \cdot n^2$?

$n^3 = n \cdot$ ______ $\cdot$ ______

$n^2 = n \cdot$ ______

$n^3 \cdot n^2 = n \cdot$ ______ $\cdot$ ______ $\cdot$ ______ $\cdot$ ______ $=$ ______

When multiplying numbers or variables with the same base, add the exponents.

Write each quotient with a common base and a single exponent.

Model 2

What is the quotient of $5^5 \div 5^3$?

$5^5 \div 5^3 = \frac{5^5}{5^3}$

$\frac{5 \cdot 5 \cdot \cancel{5} \cdot \cancel{5} \cdot \cancel{5}}{\cancel{5} \cdot \cancel{5} \cdot \cancel{5}} =$

$5 \cdot$ ______ $=$ ______

What is the quotient of $n^4 \div n$?

$n^4 \div n = \frac{n^4}{n^1}$

$\frac{n \cdot n \cdot n \cdot \cancel{n}}{\cancel{n}} =$

$n \cdot n \cdot$ ______ $=$ ______

When dividing numbers or variables with the same base, subtract the exponents.

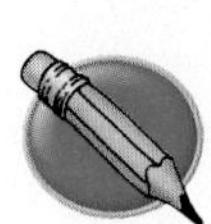

What will be the result when x^4 is divided by x^4? Explain your answer.

____

____

____

Practice

Write each product with a common base and a single exponent.

1. $3^2 \cdot 3^5 =$

   $3 \cdot 3 \cdot 3 \cdot 3 \cdot 3 \cdot 3 \cdot 3 = 3^7$

2. $10^1 \cdot 10^5 =$

3. $2^2 \cdot 2^3 =$

4. $5^2 \cdot 5^5 \cdot 5^2 =$

5. $a^4 \cdot a^3 =$

6. $y^1 \cdot y^3 \cdot y^3 =$

Write each quotient with a common base and a single exponent.

7. $24^5 \div 24^3 =$

   $\frac{24 \cdot 24 \cdot \cancel{24} \cdot \cancel{24} \cdot \cancel{24}}{\cancel{24} \cdot \cancel{24} \cdot \cancel{24}} = 24^2$

8. $15^3 \div 15^1 =$

9. $b^8 \div b^2 =$

10. $w^7 \div w^6$

Use the rules of exponents to simplify the following.

11. $2a \cdot a^3 =$

    $2 \cdot a \cdot a \cdot a \cdot a = 2a^4$

12. $8y^5 \div 2y^3 =$

Write a summary of the rules for multiplying and dividing with exponents.

____

____

____

____

Lesson 8 Scientific Notation

Very large numbers with zeros can be written as smaller numbers multiplied by multiples of 10. **Scientific notation** is used to write large numbers using powers of 10.

Model 1

To write 2,800,000,000 in scientific notation, first write it as the product of a number less than 10 and a multiple of 10.

$$2{,}800{,}000{,}000 = 2.8 \cdot 1{,}000{,}000{,}000 =$$

$$2.8 \cdot (10 \cdot 10 \cdot 10 \cdot 10 \cdot 10 \cdot 10 \cdot 10 \cdot 10 \cdot 10) = 2.8 \cdot \underline{\hspace{2cm}}$$

Notice that the decimal point was moved __________ places to write 2.8. This is the same as the ____________________ of 10 when the number is written in scientific notation.

Scientific notation makes it easier to multiply or divide large numbers.

Model 2

Multiply 130,000 · 24,000. Rewrite the factors using scientific notation.

$130{,}000 = 1.3 \cdot$ ______ and $24{,}000 =$ ______ $\cdot 10^4$

Now multiply:

$1.3 \cdot$ ______ $\cdot$ ______ $\cdot 10^4 =$

$(1.3 \cdot$ ______$) \cdot$ ______ $\cdot 10^4 =$ ______ $\cdot$ ______

Model 3

Divide 48,000 ÷ 120. Rewrite the problem using scientific notation.

$48{,}000 = 4.8 \cdot$ __________ and $120 = 1.2 \cdot$ __________

$\frac{4.8 \cdot 10^4}{1.2 \cdot 10^2} = 4.0 \cdot$ __________

Why is a base of 10 used in scientific notation?

____

____

____

Practice

Write each number in scientific notation.

1. 621,000,000 =

$6.21 \cdot 10^8$

2. 808,000,000,000 =

3. 4,900 =

4. 72,300,000 =

Multiply. Show your work.

5. 4,000,000 · 15,000 =

$4.0 \cdot 10^6 \cdot 1.5 \cdot 10^4 = 6.0 \cdot 10^{10}$

6. 300 · 522,200 =

7. 29,000,000 · 100 =

8. 8,020,000 · 7,500 =

Divide. Show your work.

9. 500,000,000 ÷ 2,000 =

$\frac{5.0 \cdot 10^8}{2.0 \cdot 10^3} = 2.5 \cdot 10^5$

10. 140,000 ÷ 14,000

11. 4,400,000 ÷ 11,000 =

12. 81,000,000 ÷ 90

Astronomers, economists, and geographers could all use scientific notation in their work. Name a type of information that each one might describe using scientific notation.

Lesson 9

Multiplication of Monomials

The rules for multiplying with exponents can be expanded to include all monomials.

Use the rules of exponents to simplify expressions with monomials.

Model 1

What is the product of $5a^2 \cdot {}^{-}3ab^3$?

$5a^2 \cdot {}^{-}3ab^3 =$

$5 \cdot ({}^{-}3) \cdot a \cdot a \cdot a \cdot b \cdot b \cdot b =$

_____ $\cdot$ _____ $\cdot$ _____ $=$ _______

Model 2

Simplify $(cd)^2$.

$(cd)^2 =$

$(cd) \cdot (cd) =$

$c \cdot c \cdot d \cdot d =$

_____ $\cdot$ _____ $=$ _______

Simplify $3a\,(2b)^3$.

$3a\,(2b)^3 =$

$3 \cdot a \cdot (2b) \cdot (2b) \cdot (2b) =$

$3 \cdot 2 \cdot$ _____ $\cdot$ _____ $\cdot\ a \cdot$ _____ $\cdot$ _____ $\cdot$ _____ $=$

_____ $\cdot\ a \cdot b^3 =$ _______

Model 3

What is the meaning of $(n^4)^2$? Use the rules of exponents to simplify.

$(n^4)^2 =$

$(n^4) \cdot (n^4) =$

$n \cdot n \cdot n \cdot n \cdot$ _____ $\cdot$ _____ $\cdot$ _____ $\cdot$ _____ $=$ _____

When raising an integer power to an integer power, multiply the exponents.

Does $8x^2 = (8x)^2$? Explain your answer.

____

____

____

Practice

Find the product.

1. $p^2q \cdot pq^3 =$

p^3q^4

2. $^-5de^4 \cdot {}^-2df =$

3. $6tu \cdot 4t^2u^3 =$

4. $3a^2b \cdot 3c^2d =$

Simplify.

5. $(3a)^3 =$

6. $(^-3stu)^4$

7. $5(np)^2 =$

8. $(2m)^4 \cdot 3m$

9. $7a^2(b^2) =$

10. $(5jk)^2 =$

Simplify.

11. $(y^3)^5 =$

y^{15}

12. $(x^7)^4 =$

13. $(p^2qr)^2$

14. $^-2(y^4)^3$

Explain how to use the rules for multiplying exponents to simplify $(ab^2)^3$.

____

____

Division of Monomials

The rules for dividing numbers with exponents also apply to dividing monomials.

Model 1

What is the quotient $\frac{r^3s^4t}{rs^2}$, when $r \neq 0, s \neq 0$?

$$\frac{r^3s^4t}{rs^2} = \frac{r \cdot r \cdot \cancel{r} \cdot s \cdot s \cdot \cancel{s} \cdot \cancel{s} \cdot t}{\cancel{r} \cdot \cancel{s} \cdot \cancel{s}} = \frac{r \cdot r \cdot s \cdot s \cdot t}{1} = r^2s^2t$$

Note that in division, the exponent of a variable in the quotient is the difference between the exponent in the numerator and the exponent in the denominator.

When expressions in the quotient contain coefficients, simplify the coefficients just like any rational number. Follow the rules for exponents for the variables.

Model 2

Divide $\frac{^-3a^5(4b)}{^-2a^2b^2}$ when $a \neq 0, b \neq 0$.

$$\frac{^-3a^5(4b)}{^-2a^2b^2} = \frac{^-3 \cdot 4 \cdot a \cdot a \cdot a \cdot a \cdot a \cdot b}{^-2 \cdot a \cdot a \cdot b \cdot b} = \frac{3 \cdot 2 \cdot a \cdot a \cdot a}{b} = \_\_\_\_\_\_\_\_$$

Sometimes it is necessary to use a combination of multiplication and division rules.

Model 3

Divide $\frac{4(2b^3c)^2}{b^2c}$ when $b \neq 0, c \neq 0$.

$$\frac{4(2b^3c)^2}{b^2c} = \frac{4 \cdot (2b^3c)(2b^3c)}{b^2c} = \frac{4 \cdot 2 \cdot 2 \cdot b \cdot b \cdot b \cdot b \cdot b \cdot b \cdot c \cdot c}{b \cdot b \cdot c} =$$

$$\_\_\_\_\_\_\_\_ = \_\_\_\_\_\_\_\_$$

Remember that an exponent outside parentheses means to use everything inside the parentheses as factors.

Why it is sometimes necessary to state that a variable cannot equal zero?

____

____

____

Practice

Find the quotient. Assume the denominator does not equal zero.

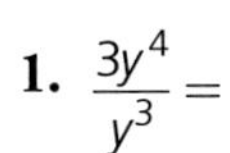

1. $\frac{3y^4}{y^3} =$

   $3y$

2. $\frac{6w^3z}{w^3} =$

3. $\frac{24a^3b^3}{^-4b^3} =$

4. $\frac{^-5x^2y^3}{10xy} =$

5. $\frac{18w^8y^4}{^-9wy^3} =$

6. $\frac{14d^5n^2}{21d^3n^2} =$

7. $\frac{32n^4p^2}{^-32np} =$

8. $\frac{4a^4b^2c}{6bc} =$

9. $\frac{^-d^7g^4h^5}{d^3gh^2} =$

Find the quotient. Assume the denominator does not equal zero.

10. $\frac{3(n^4)^2}{n^5} = \frac{3n^8}{n^5} =$

    $3n^3$

11. $\frac{2a(b^2)^2}{ab} =$

12. $\frac{(w^4y)^3}{(wy)^2} =$

13. $\frac{^-3(ab^3)^2}{6a^2b} =$

14. $\frac{(2a)^3(3b^2)^2}{4ab} =$

15. $\frac{^-5n^2(w^2)^4}{15wn^2} =$

For the problem $\frac{10p^2q^4}{2q}$, a student gave the quotient as $8p^2q^4$.

Describe the errors the student made.

____

____

Simplified Expressions

Remember that like terms have the same variables with the same exponents.

Complete the following.

Model 1

Expression	Operation	Simplified expression
$5a - 2a$	Subtract like terms.	$(5 - 2)a =$ ______
$a^2 + (^-6a^2)$	Add like terms.	$(1 + {}^-6)a^2 =$ ______

To simplify an expression, combine the coefficients of like terms.

Sometimes, more than one step is necessary to simplify an expression.

Complete the following.

Model 2

Expression	Operation	Simplified expression
$(8b)(b^3) + 2b^4$	Multiply like bases; then add like terms.	______ $+ 2b^4 =$ ______
$9a^5 - (3a^4)(^-a)$	Multiply like bases; then subtract like terms.	$9a^5 -$ (______) $=$ $9a^5 +$ ______ $=$ ______

An expression may not always simplify to a single monomial.

Model 3

Expression	Operation	Simplified expression
$(4c - c) + 6cd$	Subtract like terms.	(________)$c + 6cd =$ ______ $+ 6cd$
$^-3c + 4c^2 + 5c + 3d$	Add like terms.	$4c^2 +$ (________)$c + 3d =$ $4c^2 +$ ______$c + 3d$

Is $3x^2 - 2xy + y - 7$ in simplest form? Explain your answer.

____

____

Practice

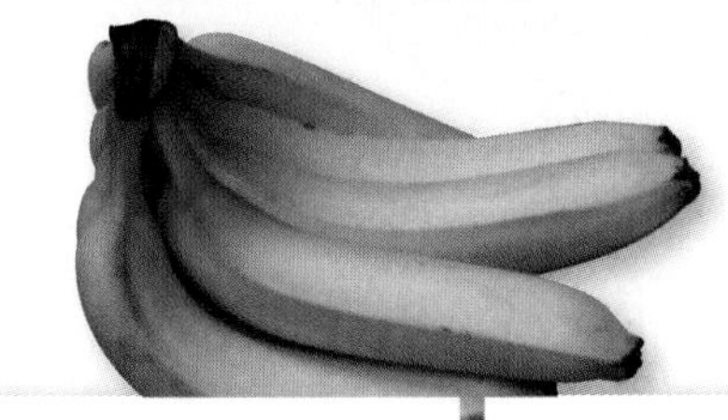

Simplify each expression.

1. $3p + 5p =$

   $8p$

2. $8rs^2 - rs^2 =$

3. $^{-}2q + 4q + (^{-}6q) =$

4. $9w^5u - 14w^5u =$

5. $^{-}7a^3 + 9a^3 =$

6. $11b - 4b - 7b =$

7. $6y - (^{-}4y) =$

8. $10ab + 10ab =$

9. $21c^2 - 32c^2 =$

Simplify each expression.

10. $(d^2e)(e^2d) + d^3e^3 =$

    $2d^3e^3$

11. $8b^5 - (2b^4)(^{-}b) =$

12. $^{-}a^2 - (3a)(a) =$

13. $10jk^3 - (2j)(4k^3) =$

14. $(n^4)(^{-}4n) + 2n^5 =$

15. $(cd)(2c) + (2c^2)d =$

Simplify each expression.

16. $3a + 8 + a =$

    $4a + 8$

17. $5w^5 - (4v^4)(2v) =$

18. $12p - 2p^2 + 3p^2 =$

Write an expression with at least three terms and two operations that can be simplified to $3a^2$.

____

____

Lesson 12 Negative Exponents

Study the table below to see how fractions and negative exponents are related.

Number	8	4	2	1	$\frac{1}{2}$	$\frac{1}{4}$	$\frac{1}{8}$
Power of 2	2^3	2^2	2^1	2^0	2^{-1}	2^{-2}	2^{-3}
Expanded form	$2 \cdot 2 \cdot 2$	$2 \cdot 2$	2	$\frac{1}{1}$	$\frac{1}{2}$	$\frac{1}{2 \cdot 2}$	$\frac{1}{2 \cdot 2 \cdot 2}$

The rule for the pattern of numbers is to divide by __________.

The pattern for the exponents is to ____________________.

Notice: $2^1 = 2$ $\quad 2^3 = 8$

$2^{-1} = \frac{1}{2}$ $\quad 2^{-3} = \frac{1}{8}$

The **reciprocal** of 2^1 is 2^{-1}. The reciprocal of 2^{-2} is __________.

Write a number with a negative exponent as the reciprocal of the number with a positive exponent.

Model 1

3^{-4} is the reciprocal of 3^4.

$3^{-4} = \frac{1}{3^4}$

2^{-5} is the reciprocal of 2^5.

$2^{-5} =$ __________

A variable with a negative exponent is the reciprocal of the variable with a positive exponent.

Model 2

b^{-6} is the reciprocal of b^6.

$b^{-6} = \frac{1}{\_\_\_\_\_\_}$

$(xy)^{-1}$ is the reciprocal of $(xy)^1$.

$(xy)^{-1} = \frac{1}{\_\_\_\_\_\_}$

Describe how to use the rules for adding and subtracting like terms when the exponents are negative.

____

____

____

____

Practice

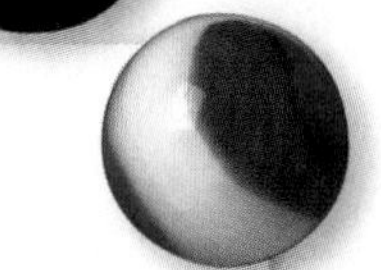

Express as a fraction with a positive exponent.

1. 10^{-8} $\frac{1}{10^8}$	**2.** 3^{-3}	**3.** $(^-4)^{-2}$
4. 9^{-2}	**5.** $^-7^{-3}$	**6.** 49^{-2}
7. 11^{-1}	**8.** 16^{-4}	**9.** 8^{-3}

Express as a fraction with a positive exponent.

10. c^{-5} $\frac{1}{c^5}$	**11.** n^{-3}	**12.** $^-4w^{-5}$
13. $2(pq)^{-4}$	**14.** $^-b^{-3}$	**15.** $(wy)^{-2}$
16. $a(b^{-4})$	**17.** $(3n)^{-2}$	**18.** $(cd)^{-6}$

How would you simplify $\frac{n^{-2}}{1}$? Explain your answer.

____

____

Lesson 13

Rational Expressions

A **rational expression** is an algebraic expression with a numerator and a denominator. Simplifying rational expressions is similar to simplifying fractions.

Complete the table. Assume no denominator equals zero.

Model 1

Expression	Reduce common factors	Simplified expression
$\frac{4}{6}$	$\frac{2 \cdot 2}{2 \cdot 3}$	______
$\frac{5a + 10}{(a + 2)}$	$\frac{5(a + 2)}{(a + 2)}$	5
$\frac{a^4(b + 2)}{a^2(b + 2)^2}$	$\frac{a^4(b + 2)}{a^2(b + 2)(b + 2)}$	______
$\frac{(a - 1)^3}{(a - 1)}$	$\frac{(a - 1)(a - 1)(a - 1)}{(a - 1)}$	______

In a simplified rational expression the numerator and denominator have no common factor except 1. Sometimes, more than one step is necessary to simplify a rational expression.

Complete the table.

Model 2

Expression	Simplified expression
$\frac{(ab)^2}{2a^2b^2c}$	$\frac{(ab)(ab)}{2a \cdot a \cdot b \cdot b \cdot c} =$ ______
$\frac{(a + 3)^3}{(2a - a + 3)}$	$\frac{(a + 3)(a + 3)(a + 3)}{(a + 3)} =$ ______

Does $\frac{-x^3}{-x^3} = \frac{x^{-3}}{x^{-3}}$? Show your work.

____

____

____

____

____

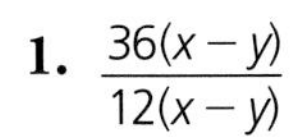

Simplify. Assume no denominator equals zero.

1. $\frac{36(x-y)}{12(x-y)}$

   3

2. $\frac{3(g^2+1)^2}{4(g^2+1)}$

3. $\frac{p^2(p+1)}{p^4}$

4. $\frac{(j+k)(j-k)}{(j-k)^2}$

5. $\frac{^-4c(c+3)}{8c^2}$

6. $\frac{-n^2}{n}$

Simplify. Assume no denominator equals zero.

7. $\frac{(5n+6+2n)}{(7n+6)^2}$

   $\frac{1}{(7n+6)}$

8. $\frac{4(pq)^3(a-b)}{(a-b)p^3}$

9. $\frac{(x+1)^2}{(x+1)(x-2)}$

10. $\frac{5p^2q(4r)^2}{10pq^2r^2}$

Explain how to solve $\frac{3(x+2)(x+3)}{(x+2)} = 0$ for x, then solve.

___

___

Expressions

This game is called *Expressions*. You can play with two or more people or teams.

Create the Game Cards

Use 20 index cards or pieces of paper to make a set of expression cards. Make the cards as follows:

ab	$\frac{a}{b}$	$a - b$	$ab + c$
$ab - c$	abc	$\frac{ab}{c}$	$a + bc$
$a - bc$	$\frac{a}{b} + c$	$\frac{a}{b} - c$	$\frac{a + b}{c}$
$\frac{(a - b)}{c}$	$(a + b)c$	$(a - b)c$	$(ab)c$
$\left(\frac{a}{b}\right)c$	$a + b$	$\frac{ab}{a}$	Wild Card

How to Play *Expressions*

The first player draws a card and puts it face up on the table. All the players then have 60 seconds to make the greatest possible number by substituting an integer value between $^-5$ and 5 for each variable on the card. The player whose expression has the greatest value gets one point.

If a player draws the wild card, the player must make up an expression using two or three variables. All players play the round using the made up expression.

There may be more than one way to create the greatest number. In the case of a tie, both players score one point.

Scoring

Use a score sheet like the one shown below to keep track of each round. The player or players who make the largest number in a round get a point. The player with the most points after 20 rounds of the game wins.

Round	Expression	Carrie	pts	Daniel	pts	LaQuita	pts
1	$a + b$	$5 + 5 = 10$	1	$5 + {}^-5 = 0$	0	${}^-5 + {}^-5 = {}^-10$	0
2	$a - b$	${}^-5 - ({}^-5) = 0$	0	$5 - ({}^-5) = 10$	1	$5 - ({}^-5) = 10$	1
3	abc	$5 \cdot 5 \cdot 5 = 125$	1	${}^-5 \cdot {}^-5 \cdot 5 = 125$	1	$5 \cdot 5 \cdot 5 = 125$	1
20							
Total Points:		Carrie		Daniel		LaQuita	

Variations

A. Players must make the least number possible.

B. Add additional expression cards, such as a^{b-c}.

C. Make a set of five operator cards (+, −, ·, ÷, exponent). Draw two expression cards and join them with an operator card.

Unit 1 Review

Find the value.

1. $10 \cdot {}^-36 =$
2. ${}^-82 \cdot {}^-3 =$
3. $3^4 \cdot 0 =$
4. $({}^-10)^5 =$
5. ${}^-4 - 8 =$
6. ${}^-39 \div {}^-13 =$
7. $33 - ({}^-72) =$
8. $2^8 \div 2^3 =$
9. $5^2 \cdot 1 =$
10. $15 + ({}^-24) =$

Find each product or quotient. Write the answer in scientific notation.

11. $9{,}600{,}000 \div 3{,}200 =$
12. $58{,}000 \cdot 600{,}000 =$

Express with positive exponents.

13. $(e^{-4})(e^5y^3)$ $\quad e \neq 0$
14. $\frac{9r^6}{3r^3s^2}$ $\quad r \neq 0, s \neq 0$

Simplify.

15. ${}^-3a^3 + ({}^-10a) + 2a$
16. $6p^2r^3 - ({}^-9p^2r^3)$
17. $(5ab^3)({}^-9b^9)$
18. $\frac{{}^-18w^9x}{{}^-6w^2x}$ $\quad w \neq 0, x \neq 0$
19. $(7d^4g)^2$
20. $4c(2n)^3$
21. $\frac{(4ab^3)^3}{2ab}$ $\quad a \neq 0, b \neq 0$
22. $\frac{10(a+b)^2}{2(a+b)}$ $\quad a \neq 0, b \neq 0$

Properties and Equations

Unit 2

Oooh! Aaah!

Do you enjoy a good fireworks show?

Did you know that an 8-inch diameter fireworks shell rises to about 525 feet and creates a burst of 360 feet in diameter? People who create fireworks calculate different sizes of bursts and different heights. What math do you suppose they use to compare the diameter of a shell and the diameter of a burst?

Lesson 1 Properties of Addition

The properties of addition for numbers apply to algebraic expressions.

- The **Associative Property of Addition** allows regrouping of terms.
- The **Commutative Property of Addition** allows a change in the order of terms.
- The **Identity Property of Addition** states that the addition of zero to a term does not change its value.
- The **Inverse Property of Addition** states that the sum of a term and its opposite is zero.

Property	Definition	Example
Associative Property of Addition	$(a + b) + c = a + (b + c)$	$(3 + 4) + 7 = 3 + (4 + 7)$ $7 + 7 = 3 + 11$ $14 = 14$
Commutative Property of Addition	$a + b = b + a$	$4 + 6 = 6 + 4$ $10 = 10$
Identity Property of Addition	$a + 0 = a$	$8 + 0 = 8$ $8 = 8$
Inverse Property of Addition	$a + (^-a) = 0$	$9 + (^-9) = 0$ $0 = 0$

Complete the following. State the property of addition used.

Model 1

$2d + (7d + 3d) = (2d + 7d) +$ ______ Associative Property

$5b +$ ______ $= 5b$ ______________________

$4n + n =$ ______ $+ 4n$ ______________________

______ $+ (^-8x) = 0$ ______________________

Explain which property of addition is shown in the statement $7a + (2a + 4a) = (2a + 4a) + 7a$.

____

____

____

____

____

Practice

Apply a property of addition to fill in the blanks. State the property used.

1. $3r + (10r + 13r) = (\underline{3r} + \underline{10r}) + 13r$

   Associative Property

2. $t + 12t =$ ______ + ______

   ____

3. ______ $= 19k + (^-19k)$

   ____

4. (______ + ______) $+ q = 5q + (7q + q)$

   ____

5. $4w +$ ______ $= 4w$

   ____

6. $9v +$ ______ $= 5v +$ ______

   ____

7. $0 = 4n^2 +$ (______)

   ____

8. ______ + (______ $+ x) = (9x + 2x) + x$

   ____

9. ______ $+ (^-8y) = 0$

   ____

10. $16t^2 +$ ______ $= 16t^2$

    ____

11. $0 + (x + 4x) =$ (______ + ______) $+ 4x$

    ____

12. $(6g + (^-6g)) + 7g =$ (______ $+ 6g) + 7g$

    ____

Explain how to use the Commutative Property to show $(a + b) + (c + d) = (d + c) + (b + a)$

____

____

Properties of Multiplication

The properties of multiplication for numbers apply to algebraic expressions.

- The **Associative Property of Multiplication** states that changing the grouping of factors does not change the product.
- The **Commutative Property of Multiplication** states that changing the order of factors does not change the product.
- The **Identity Property of Multiplication** states that the product of 1 and any term is that term.
- The **Inverse Property of Multiplication** states that the product of multiplicative inverses is always 1.

For any number, *a*, the **multiplicative inverse** or **reciprocal** is $\frac{1}{a}$.

Property	Definition	Example
Associative Property of Multiplication	$(a \cdot b) \cdot c = a \cdot (b \cdot c)$	$(6 \cdot 4) \cdot 3 = 6 \cdot (4 \cdot 3)$ $24 \cdot 3 = 6 \cdot 12$ $72 = 72$
Commutative Property of Multiplication	$a \cdot b = b \cdot a$	$5 \cdot 2 = 2 \cdot 5$ $10 = 10$
Identity Property of Multiplication	$a \cdot 1 = a$	$12 \cdot 1 = 12$ $12 = 12$
Inverse Property of Multiplication	$a \cdot \frac{1}{a} = 1$, When $a \neq 0$	$8 \cdot \frac{1}{8} = 1$ $\frac{8}{8} = 1$

Complete the following. State the property of multiplication used.

Model 1

$5y \cdot 1 = 5y$ Identity Property of Multiplication

$2b \cdot 6b = 6b \cdot$ ______ __________________

$(4n \cdot 5n) \cdot 9n = 4n \cdot ($______ $\cdot 9n)$ __________________

______ $\cdot 8x = 1$ __________________

Compare the Identity Property of Addition to the Identity Property of Multiplication. How are they alike and how are they different?

____

____

____

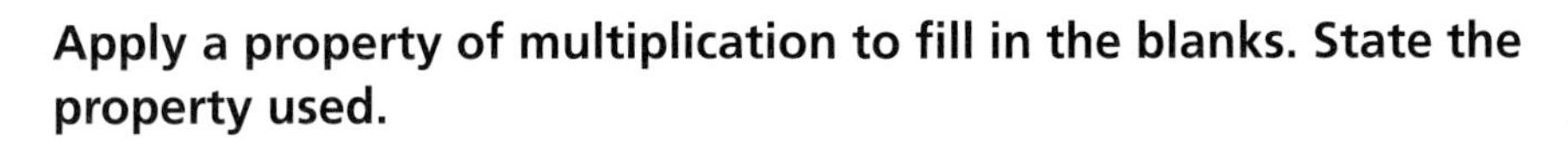

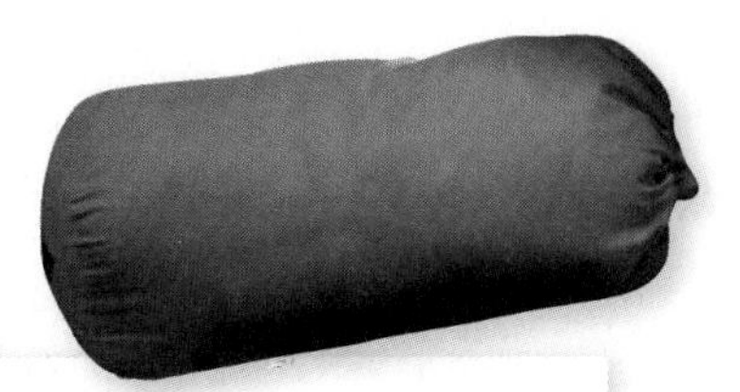

Apply a property of multiplication to fill in the blanks. State the property used.

1. $10r \cdot 7r = \underline{\quad 7r \quad} \cdot \underline{\quad 10r \quad}$

   Commutative Property

2. $\frac{n}{5} \cdot \underline{\qquad\quad} = 1$

   ____

3. $y^2 \cdot 6 = \underline{\qquad\quad} \cdot y^2$

   ____

4. $18 \cdot (2u \cdot 5v^2) = (\underline{\qquad\quad} \cdot \underline{\qquad\quad}) \cdot 5v^2$

   ____

5. $\underline{\qquad\quad} \cdot w = w$

   ____

6. $^-34cd = {}^-34\underline{\qquad\quad}$

   ____

7. $\underline{\qquad\quad} \cdot \frac{a}{b} = 1$

   ____

8. $(4g \cdot g) \cdot 5 = 4g \cdot (\underline{\qquad\quad} \cdot \underline{\qquad\quad})$

   ____

9. $3d \cdot \underline{\qquad\quad} = 3d$

   ____

10. $a \cdot \left(\frac{a}{b} \cdot d\right) = \left(\underline{\qquad\quad} \cdot \underline{\qquad\quad}\right) \cdot d$

    ____

11. $5b \cdot 8b \cdot 1 = \underline{\qquad\quad} \cdot 5b \cdot 1$

    ____

12. $\underline{\qquad\quad} \cdot 1 = \underline{\qquad\quad}$

    ____

What happens when a number or algebraic term is multiplied by zero? Using variable *a*, write a mathematical equation representing this property of zero.

____

____

Lesson 3

Order of Operations

The **Order of Operations** is a set of directions for simplifying numerical or algebraic expressions. Follow these steps to simplify expressions.

- Perform all operations in parentheses.
- Perform all operations with exponents.
- Perform all multiplication and division in order from left to right.
- Perform all addition and subtraction in order from left to right.

Follow the Order of Operations to simplify each numerical expression.

Model 1

Simplify $3 + 9 \cdot 6 \div 3$.

$3 + 9 \cdot 6 \div 3 =$

$3 +$ ________ $\div 3 =$

$3 +$ ________ $=$

____

Simplify $(2 + 8)^2 \cdot 3 + 5$.

$(2 + 8)^2 \cdot 3 + 5 =$

(________$)^2 \cdot 3 + 5 =$

$100 \cdot 3 + 5 =$

________ $+ 5 =$

____

Follow the Order of Operations to simplify each algebraic expression.

Model 2

Simplify $x + 2(^-4x) \div 8x$.

$x + 2(^-4x) \div 8x =$

$x +$ ________ $\div 8x =$

$x +$ ________ $=$

____

Simplify $(2x)^2 \cdot 4 - 5$.

$(2x)^2 \cdot 4 - 5 =$

________ $\cdot 4 - 5 =$

________ $- 5 =$

____

Suppose that two different calculators are used to compute the following: $3 \cdot 4 + 7 \cdot 2$. One calculator computes the answer as 26. The other calculator gives 38. Which answer is correct, and what could explain the difference?

____

____

____

Practice

Simplify each expression.

1. $6 \cdot 2 \div (^{-}3 + {^{-}3}) =$

   $^{-}2$

2. $5 + 12 \div 3 - 16 =$

3. $11 + (7 - 3)^2 =$

4. $1 + 2 - 3 \cdot 4 \div 6 =$

5. $2 + (11 - 13) + 8 =$

6. $20 + 4 \div (^{-}2) - 1 =$

Simplify each expression.

7. $(4t^2 + 6t^2) \cdot 8 =$

   $80t^2$

8. $r \cdot 3r - 7r^2 =$

9. $9c^2 + (2c)^2 + (5c)^2 =$

10. $9y - 3y \cdot y + 9y =$

11. $5 \cdot (3y)^2 + y^2 =$

12. $12n \cdot 3 + 3 \cdot 4n =$

When parentheses appear inside parentheses in an expression, work from the inside to the outside. Explain how to apply the order of operations to simplify the expression $(6 + (8 - 7))^2 + 7 \cdot 3$.

Distributive Property

The **Distributive Property** relates the operations of multiplication and addition. The property is helpful for simplifying both numerical and algebraic expressions.

Definition	Meaning	Example
$a(b + c) = a \cdot b + a \cdot c$	Multiply each factor in the parentheses by the factor *a*.	$9(4 + 7) =$ $9 \cdot 4 + 9 \cdot 7 =$ $36 + 63 =$ 99

Use the Distributive Property to simplify.

Model 1

$8(11 - 7) =$

$8[11 + (^{-}7)] =$

$8 \cdot 11 + 8 \cdot$ ______ $=$

$88 +$ ______ $=$

______

$7(3y + 8z) =$

$7 \cdot 3y + 7 \cdot$ ______ $=$

______ $+$ ______

$5(11 + 4 + {}^{-}1) =$

$5 \cdot 11 + 5 \cdot$ ______ $+ 5 \cdot$ ______ $=$

$55 +$ ______ $+$ ______ $=$

______

$2(5a + 8b + 3b) =$

$2 \cdot 5a + 2 \cdot$ ______ $+ 2 \cdot$ ______ $=$

$10a +$ ______ $+$ ______ $=$

$10a +$ ______

Rewrite expressions by looking for greatest common factors and applying the Distributive Property in reverse.

Model 2

$38 + 4 =$

$2 \cdot$ ______ $+ 2 \cdot$ ______ $=$

$2($______ $+$ ______$)$

$6g + 7gh =$

$g \cdot$ ______ $+ g \cdot$ ______ $=$

$g($______ $+$ ______$)$

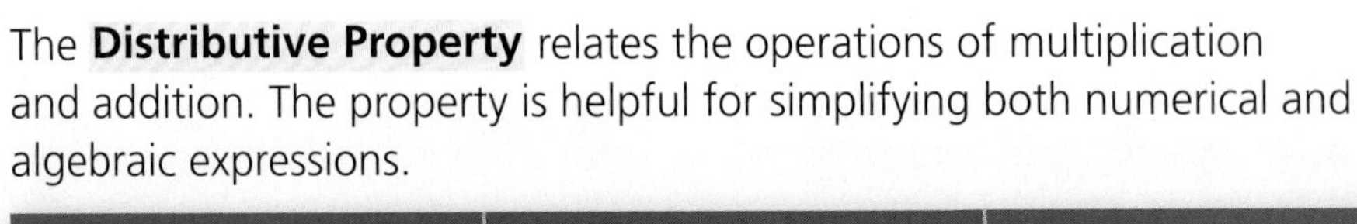

Explain how to apply the Distributive Property to the expression 9(8*a* – 5*b*).

____

____

Practice

Use the Distributive Property to simplify.

1. $3(3q - 4) =$

   $3[3q + (^{-}4)] =$

   $3 \cdot 3q + 3 \cdot (^{-}4) =$

   $9q + (^{-}12)$ or $9q - 12$

2. $5(3d + 7) =$

3. $^{-}5(2 + 3k) - 4k =$

4. $14 + 5(a + b) =$

5. $20n + 2n(1 + 4) =$

6. $^{-}8(3w + 2w) =$

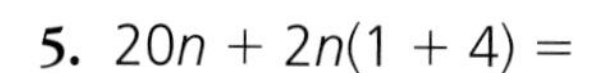

7. $2(w + 8z) + 3(w + z) =$

8. $2m(y - 3m) =$

Rewrite using the Distributive Property and the greatest common factor of each term.

9. $3a + 9a =$

   $3a(1 + 3)$

10. $21m - 12mn =$

11. $18r^2s^3 + 6r =$

12. $9 + 6n =$

13. $12b + 2y =$

14. $10a^2 + 15a =$

To simplify an expression, a student wrote $4(x + 3) + 5 = 4x + 8$. Explain the error in this statement.

____

____

Lesson 5 Addition and Subtraction Equations

An **equation** shows that two expressions are equal. To solve an equation, isolate the variable by performing inverse operations.

An equation is like a balance with equal weights. If you change the weight on one side, you must change the weight on the other side in the same way to maintain the balance.

Solve each equation.

Model 1

$^{-}2 + x = 0$

$^{-}2 + x + \_\_\_\_\_ = 0 + \_\_\_\_\_$

$\_\_\_\_\_ + x = \_\_\_\_\_$

$x = \_\_\_\_\_$

Think: How can you move all the numbers to the right side of the equation?
Add 2 to both sides.
Combine terms.

Check: $^{-}2 + 2 = 0$

$12 = x - 4$

$12 + \_\_\_\_\_ = x - 4 + \_\_\_\_\_$

$\_\_\_\_\_ = x + \_\_\_\_\_$

$\_\_\_\_\_ = x$

Think: How can you move all the numbers to the left side of the equation?
Add 4 to both sides.
Combine terms.

Check: $12 = 16 - 4$

Write an equation to model the situation. Then solve the problem.

Model 2

Hector collects postcards. His goal is to own one postcard from each of the 50 states. So far, he has postcards from 27 states.

$27 + p = 50$

$27 + p - \_\_\_\_\_ = 50 - \_\_\_\_\_$

$\_\_\_\_\_ + p = \_\_\_\_\_$

$p = \_\_\_\_\_$

Let p represent the postcards Hector needs.

Hector needs _____ postcards to complete his collection.

Describe how to check a solution to an equation.

____

____

Practice

Solve each equation. Show each step. Then check your answer.

1. $a + 24 = 100$

$a + 24 - 24 = 100 - 24$

$a = 76$

Check:

$76 + 24 = 100$

2. $h + 3 = 213$

3. $q - 14 = 71$

4. $s - 25 = 39$

5. $55 + x = 30$

6. $^{-}17 + b = {}^{-}10$

7. $14 = y + 5$

8. $26 = {}^{-}2 + n$

9. $m - 10 = {}^{-}18$

Write an equation with variable *n* to model the situation. Then solve the problem.

10. Julia is running a 5,000-meter race. She has 350 meters left to go. How far has she run?

Equation: $n + 350 = 5{,}000$

Solution: $n + 350 - 350 = 5{,}000 - 350$

$n = 4{,}650$ meters

11. Jamal bought 15 trading cards last week. He bought more cards this week and now has 29 cards. How many cards did Jamal buy this week?

Explain how the steps for solving $25 + p = 50$ and $^{-}25 + q = {}^{-}50$ are different. Then solve each equation.

Multiplication and Division Equations

Multiplication and division equations can be solved using algebra. To keep an equation balanced, perform the same operation on both sides of the equation to isolate the variable.

Model 1

$48 = 4c$

$\frac{48}{\_\_\_\_} = \frac{4c}{\_\_\_\_}$

$\_\_\_\_ = c$

Think: How can you move all the numbers to the left side of the equation? Divide both sides by 4.
Simplify fractions.
Check: $48 = 4 \cdot 12$

$\frac{h}{^-3} = {^-12}$

$\_\_\_\_ \cdot \frac{h}{^-3} = {^-12} \cdot \_\_\_\_$

$h = \_\_\_\_$

Think: How can you move all the numbers to the right side of the equation? Multiply both sides by $^-3$.
Simplify.
Check: $36 \div {^-3} = {^-12}$

Write and solve an equation to model the situation.

Model 2

Jen is taking guitar lessons. Each lesson costs the same amount. After seven lessons, she has paid \$56. How much does one lesson cost?

Let *c* represent the cost of each lesson.

Number of lessons	Operation	Cost per lesson, *c*	=	Total cost
____	____	____	=	____

$7c = 56$

$\frac{7c}{7} = \frac{56}{\_\_\_\_}$

$c = \_\_\_\_$ One lesson costs \$8.

How is dividing ^-2y by $^-2$ like multiplying ^-2y by $-\frac{1}{2}$.

____

____

____

Practice

Solve each equation. Show each step. Then check your answer.

1. $16f = 32$

   $\frac{16f}{16} = \frac{32}{16}$

   $f = 2$

   Check:

   $16(2) = 32$

2. $\frac{b}{^-2} = 46$

3. $^-24q = 24$

4. $^-15n = 90$

5. $\frac{r}{25} = 25$

6. $4w = 600$

7. $^-12 = \frac{y}{3}$

8. $\frac{b}{^-5} = 35$

9. $100 = 10z$

Write and solve an equation with a variable for each situation.

10. The city bike trail makes a 6-mile loop. Becky rode her bicycle around the trail for a total of 18 miles. How many times did Becky complete the trail?

    Equation: $6t = 18$

    Solution: $t = 3$ times

11. Alvaro gets a business loan of $5,000. The monthly loan payment is $250. How many months will it take Alvaro to pay back the loan?

Explain how to solve the equation $\frac{3x}{^-5} = 30$.

____

____

____

Multi-Step Equations

More than one step may be needed to isolate the variable when solving an equation. Perform inverse operations on both sides of the equation.

Solve each equation. Then check your answer.

Model 1

$2x - 1 = 11$

$2x - 1 + \_\_\_\_ = 11 + \_\_\_\_$

$2x = \_\_\_\_$

$\frac{2x}{2} = \frac{\_\_\_\_}{2}$

$x = \_\_\_\_$

Think: The easiest first step is to deal with $^{-}1$.

Add 1 to both sides.

Combine terms.

Divide both sides by 2.

Check: $2 \cdot 6 - 1 = 12 - 1 = 11$

$\frac{x}{4} + 5 = 29$

$\frac{x}{4} + 5 - \_\_\_\_ = 29 - \_\_\_\_$

$\frac{x}{4} = \_\_\_\_$

$4 \cdot \frac{x}{4} = 4 \cdot \_\_\_\_$

$x = \_\_\_\_$

Think: The easiest first step is to deal with $^{+}5$.

Subtract 5 from both sides.

Combine terms.

Multiply both sides by 4.

Check: $96 \div 4 + 5 =$
$24 + 5 = 29$

Model 2

$3(5 + x) = {}^{-}21$

$3 \cdot \_\_\_\_ + 3x = {}^{-}21$

$\_\_\_\_ + 3x = {}^{-}21$

$15 + 3x - \_\_\_\_ = {}^{-}21 - \_\_\_\_$

$\frac{3x}{\_\_\_\_} = \frac{}{\_\_\_\_}$

$x = \_\_\_\_$

Think: Use the Distributive Property first.

Simplify.

Subtract 15 from both sides.

Divide both sides by 3.

Check: $3(5 + {}^{-}12) =$
$15 - 36 = {}^{-}21$

Explain two different ways to solve $3(x + 7) = 18$.

____

____

____

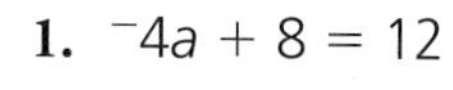

Practice

Solve each equation. Show each step. Then check your answer.

1. $^{-}4a + 8 = 12$

   $^{-}4a + 8 - 8 = 12 - 8$

   $\frac{^{-}4a}{^{-}4} = \frac{4}{^{-}4}$

   $a = {}^{-}1$

2. $7n + 4 = 32$

3. $^{-}5 + 10r = 35$

4. $23y - 27 = 88$

5. $6x - 9 = 87$

6. $2w - 18 = {}^{-}2$

7. $\frac{n}{6} + 6 = 8$

8. $7 + \frac{y}{4} = 3$

9. $17 - 4b = 9$

Solve each equation. Show each step. Then check your answer.

10. $21(t + 2) = 84$

    $21t + 42 = 84$

    $21t + 42 - 42 = 84 - 42$

    $\frac{21t}{21} = \frac{42}{21}$

    $t = 2$

11. $^{-}2(k - 5) = {}^{-}6$

12. $8(g + 6) = 24$

Describe the steps required to solve $\frac{8(6 + n) - 1}{5} = 19$.

____

____

____

____

Equations with Fractions

One way to solve equations with fractions is to find the least common multiple of the denominators. Then eliminate the fractions by multiplication.

Solve each of the following equations.

Model 1

$\frac{(x+1)}{2} = 5$ — **Think:** The least common multiple of 2 and 1 is 2.

$2 \cdot \frac{(x+1)}{2} = 2 \cdot 5$ — Multiply both sides by 2.

$(\_\_\_\_\_\_) = 10$ — Combine terms.

$x + 1 - 1 = 10 - \_\_\_\_\_\_$ — Subtract 1 from both sides.

$x = \_\_\_\_\_\_$ — **Check:** $\frac{9+1}{2} = \frac{10}{2} = 5$

Model 2

$\frac{x}{3} + \frac{2}{5} = \frac{11}{15}$ — **Think:** The least common multiple of 3 and 5 is 15.

$\_\_\_\_\_\_ \cdot \left(\frac{x}{3} + \frac{2}{5}\right) = \_\_\_\_\_\_ \cdot \frac{11}{15}$ — Multiply both sides by 15.

$15\left(\frac{x}{3}\right) + 15\left(\frac{2}{5}\right) = \_\_\_\_\_\_$ — Use the Distributive Property.

$5x + 6 = 11$

$\_\_\_\_\_\_ + 6 - 6 = \_\_\_\_\_\_ - 6$ — Subtract 6 from both sides.

$\frac{5x}{5} = \frac{5}{5}$ — Divide both sides by 5.

$x = \_\_\_\_\_\_$ — **Check:** $\frac{1}{3} + \frac{2}{5} = \frac{5}{15} + \frac{6}{15} = \frac{11}{15}$

How would you simplify the equation 3.5x + 2.75 = 10 to make it easier to solve?

____

____

Practice

Solve each equation. Show each step. Then check your answer.

1. $\frac{9 + w}{4} = 5$

$4 \cdot \frac{9 + w}{4} = 4 \cdot 5$

$9 + w = 20$

$w = 11$

2. $8 + \frac{3}{10}z = 14$

3. $\frac{d}{7} - \frac{3}{7} = 3$

4. $\frac{-5y}{6} - 2 = {}^{-}12$

5. $8 + \frac{c}{3} = 11$

6. $\frac{p}{2} + \frac{1}{2} = 18$

Solve each equation. Show each step. Check your answer.

7. $\frac{a}{27} + 1 = \frac{8}{9}$

$27 \cdot \left(\frac{a}{27} + 1\right) = 27 \cdot \frac{8}{9}$

$a + 27 = 24$

$a = {}^{-}3$

8. $\frac{4b}{9} + 4 = 0$

9. $\frac{(5 - d)}{35} = \frac{4}{7}$

10. $\frac{3}{6} + \frac{t}{4} = -\frac{1}{2}$

11. $\frac{(x + 1)}{4} = \frac{1}{2}$

12. $\frac{1}{4} + \frac{n}{8} = \frac{1}{2}$

Explain how multiplying both sides of an equation by the least common multiple can make the equation easier to solve.

Lesson 9 Equations with Like Terms

Combine like terms when they are on the same side of an equation before continuing to solve the equation.

Solve each equation.

Model 1

$n + 2n = 6$

$3n = 6$

$\frac{3n}{3} = \frac{6}{3}$

$n =$ ______

Think: Combine like terms on the same side of the equation.

Divide both sides by 3.

Check: $2 + 2(2) = 2 + 4 = 6$

Model 2

$47 = 6w - 5 - 4w$

$47 = 2w - 5$

$47 + 5 = 2w - 5 +$ ______

$52 = 2w$

$\frac{52}{2} = \frac{2w}{2}$

______ $= w$

Think: Combine like terms on the same side of the equation.

Add 5 to both sides.

Combine terms.

Divide both sides by 2.

Check: $6(26) - 5 - 4(26) =$
$156 - 5 - 104 = 47$

Model 3

$8m + 6 - \frac{m}{2} = 36$

______ $\cdot \left(8m + 6 - \frac{m}{2}\right) =$ ______ $\cdot 36$

$2(8m) + 2(6) - 2\left(\frac{m}{2}\right) =$ ______

$16m + 12 - m =$ ______

______ $+ 12 =$ ______

______ $+ 12 - 12 =$ ______ $- 12$

$m =$ ______

Think: First multiply both sides by 2.

Use the Distributive Property.

Combine like terms on the same side of the equation.

Subtract 12 from both sides.

Check: $8(4) + 6 - (4 \div 2) =$
$32 + 6 - 2 = 36$

Describe the first step in solving the equation $4n + 3 + n = 23$. Then solve for n.

______

Practice

Solve each equation. Show each step. Check your answer.

1. $8y - 3y = 25$
   $5y = 25$
   $y = 5;$
   $8(5) - 3(5) =$
   $40 - 15 = 25$

2. $5t + 8t - 10t = {}^-42$

3. $4g - g = 24$

4. $15a - 4a + 22 = 0$

5. ${}^-p + 6p = 35$

6. $48 = 10w - 7w$

Solve each equation. Show each step. Check your answer.

7. ${}^-24 = {}^-5n + 6 - 10n$
   ${}^-24 = {}^-15n + 6$
   ${}^-30 = {}^-15n$
   $n = 2;$
   ${}^-5(2) + 6 - 10(2) = {}^-24$

8. $12 - y = 27$

9. $25c - 10c - 9 = {}^-39$

10. $12z + 8 + z = {}^-18$

11. ${}^-13 + 5n - 3n = 27$

12. $4(k + 1) + 2k = {}^-2$

Solve each equation. Show each step. Check your answer.

13. $\frac{(q + 2q)}{2} = 15$
    $3q = 30$
    $q = 10;$
    $(10 + 20) \div 2 =$
    $30 \div 2 = 15$

14. $\frac{x}{4} - 12 + \frac{7x}{4} = 28$

15. $8t - 5t + 3 = {}^-18$

Describe how to solve $3a = a + 10$.

____

____

Lesson 10

Equations with Variables on Both Sides

Sometimes an equation has the same variable on both sides of the equal sign. To solve the equation, use the inverse operation to remove the variable from one side of the equation.

Solve each of the following equations.

Model 1

$5a = 2a - 6$

$5a - 2a = 2a - 6 - 2a$

$\_\_\_\_\_\_ = {}^{-}6$

$\frac{}{3} = \frac{{}^{-}6}{3}$

$a = \_\_\_\_\_\_$

Think: Both terms with a variable should be on the same side.
Subtract $2a$ from both sides of the equation.
Combine like terms.
Divide both sides by 3.

Check: $5({}^{-}2) = {}^{-}10$;
$2({}^{-}2) - 6 = {}^{-}10$

$a + 5 = 4a - 1$

$a + 5 - a = 4a - 1 - a$

$5 = \_\_\_\_\_\_ - 1$

$5 + \_\_\_\_\_\_ = \_\_\_\_\_\_ - 1 + \_\_\_\_\_\_$

$\_\_\_\_\_\_ = \_\_\_\_\_\_$

$a = \_\_\_\_\_\_$

Think: Subtract the lowest value of the coefficient first.
Subtract a from both sides.
Combine like terms.
Add 1 to both sides.
Divide both sides by 3.

Check: $2 + 5 = 7$;
$4(2) - 1 = 7$

In some cases, there is no solution to an equation, and in some cases any number will satisfy an equation.

Model 2

$6n + 9 = 6n$

$6n - 6n + 9 = 6n - 6n$

$\_\_\_\_\_\_ + 9 = 0$

$\mathbf{9 \neq 0}$

Subtract $6n$ from both sides. Combine like terms.

The statement $\mathbf{9 \neq 0}$ means that there is no solution to the equation.

Any number will satisfy the equation $2 + 3x = x + 2x + 2$. Explain why this is true.

____

____

Practice

Solve each equation. Show each step. Check your answer.

1. $3n + 6 = n$

   $3n + 6 = n$

   $6 = {}^{-}2n$

   ${}^{-}3 = n;$

   $3 \cdot {}^{-}3 + 6 = {}^{-}3$

   ${}^{-}3 = {}^{-}3$

2. ${}^{-}x - 7 = 1$

3. ${}^{-}4w = 3w - 14$

4. $17 + 4q = {}^{-}13q$

5. $4b + 3 = {}^{-}2b$

6. $\frac{3n}{2} + 5 = \frac{n}{2}$

Solve and check.

7. ${}^{-}2d + 7 = 4d - 5$

   ${}^{-}2d + 7 + 2d = 4d - 5 + 2d$

   $7 = 6d - 5$

   $12 = 6d$

   $2 = d;$

   ${}^{-}2 \cdot 2 + 7 = 4 \cdot 2 - 5 =$

   ${}^{-}4 + 7 = 8 - 5; 3 = 3$

8. $1 - 6k = k - 20$

9. $3c + 12 = 3c + 6 + 6$

10. $10t - 4 = 9t + 1$

11. $3x - 2 = {}^{-}1 + 3x$

12. $7d + 9 = 2d - 1$

Explain why there is no solution to the equation $8 + 3w = 3w$.

____

____

More Equations with Variables on Both Sides

Sometimes an equation with the variable on both sides also has fractions. Multiply to get rid of the fractions. Then gather the variables on one side.

Solve each of the following equations.

Model 1

$\frac{n}{2} = 3n - 1$	**Think:** What is the denominator?
$\_\_\_\_ \cdot \frac{n}{2} = \_\_\_\_ \cdot (3n - 1)$	Multiply both sides by 2.
$n = \_\_\_\_ - 2$	Use the Distributive Property.
$n - 6n = 6n - 6n - 2$	Subtract $6n$ from both sides.
$\_\_\_\_ = {}^{-}2$	Combine terms.
$\frac{{}^{-}5n}{{}^{-}5} = \frac{{}^{-}2}{{}^{-}5}$	Divide both sides by ${}^{-}5$.
$n = \_\_\_\_$	**Check:** $\frac{2}{5} \div 2 = \frac{1}{5}$; $3 \cdot \frac{2}{5} - 1 = \frac{1}{5}$

Model 2

$\frac{x}{2} - \frac{3}{4} = \frac{x}{4}$	**Think:** The least common multiple of the denominators is 4.
$\_\_\_\_ \cdot \left(\frac{x}{2} - \frac{3}{4}\right) = \_\_\_\_ \cdot \frac{x}{4}$	Multiply both sides by 4.
$2x - 3 = \_\_\_\_$	Use the Distributive Property.
$2x - x - 3 = 0$	Subtract x from both sides.
$\_\_\_\_ - 3 + 3 = \_\_\_\_$	Add 3 to both sides.
$x = \_\_\_\_$	**Check:** $\frac{3}{2} - \frac{3}{4} = \frac{6}{4} - \frac{3}{4} = \frac{3}{4}$

Explain the steps to solve the equation $2x - 5 + \frac{3x}{4} = \frac{x}{2} + 4$.

____

____

____

Practice

Solve each equation. Show each step. Then check your answer.

1. $\frac{n}{3} = 2n - 10$

$3 \cdot \frac{n}{3} = 3 \cdot (2n - 10)$

$n = 6n - 30$

$^{-}5n = {}^{-}30$

$n = 6$

2. $\frac{2a}{3} - 5 = a$

3. $\frac{2}{5} + 6n = 5n$

4. $2n + 7 = \frac{3n}{5}$

5. $\frac{y}{10} + 19 = 2y$

6. $\frac{2}{3} + b = {}^{-}4$

Solve each equation. Show each step. Then check your answer.

7. $\frac{x}{3} + \frac{1}{6} = x$

$6 \cdot \left(\frac{x}{3} + \frac{1}{6}\right) = 6 \cdot x$

$2x + 1 = 6x$

$1 = 4x$

$x = \frac{1}{4}$

8. $\frac{7}{2} + 3w = \frac{9}{2} + 4w$

9. $\frac{-g}{4} + 1 = g - \frac{3}{8}$

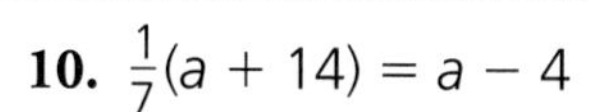

10. $\frac{1}{7}(a + 14) = a - 4$

Describes what happens when gathering the variables to one side when solving $\frac{7}{2} + 3w = \frac{9}{2} + 3w$.

Lesson 12 Addition and Subtraction Inequalities

An **inequality** shows that two expressions are not equal and describes the relationship between the expressions.

Examples: $a < 3$, read *a is less than 3* $b > 3$, read *b is greater than 3*

For each of the inequalities above, there is more than one solution. These solutions can be modeled on a number line.

Graph each inequality. Open circles mean that the number is not included in the solution. Solid circles mean the number is included in the solution set.

Model 1

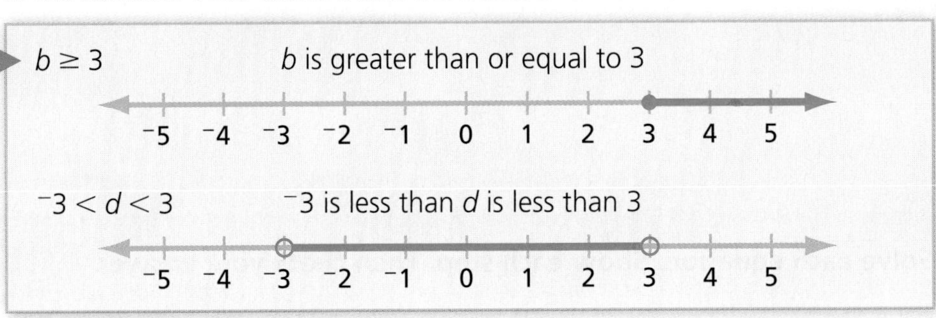

Inequalities can be solved using the same steps as equations. **Solve each inequality and graph the solution.**

Model 2

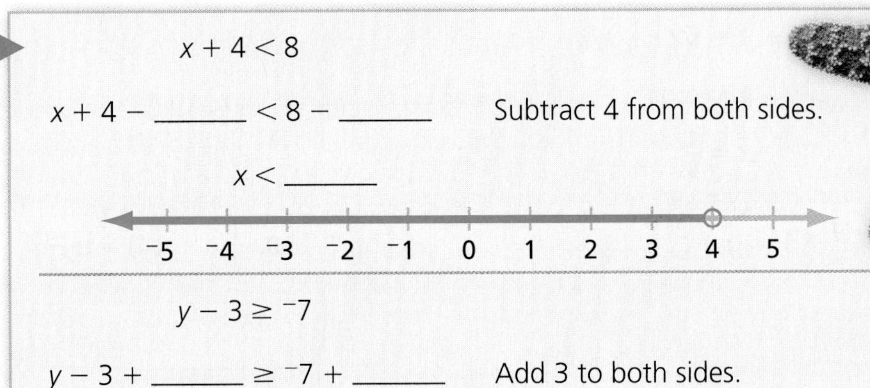

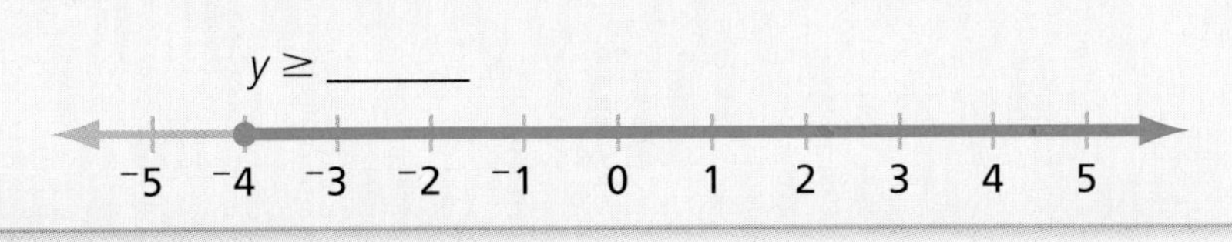

Explain how you would graph $c \neq 3$ on a number line.

____

____

____

Practice

Graph each inequality.

1. 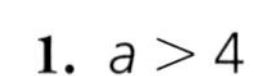

$a > 4$

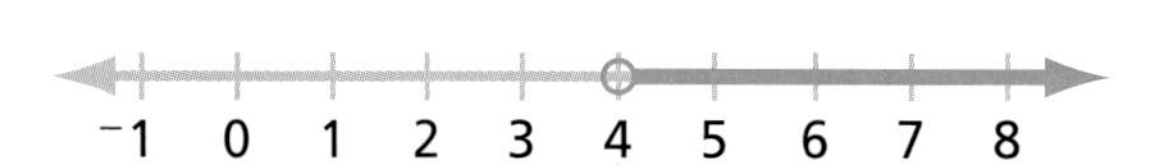

2. 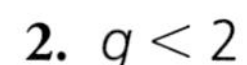

$g < 2$

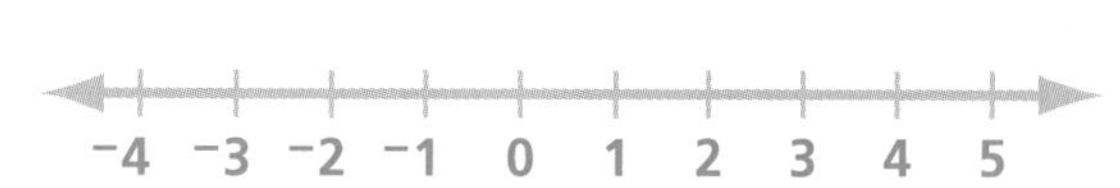

3. $2 \leq n$

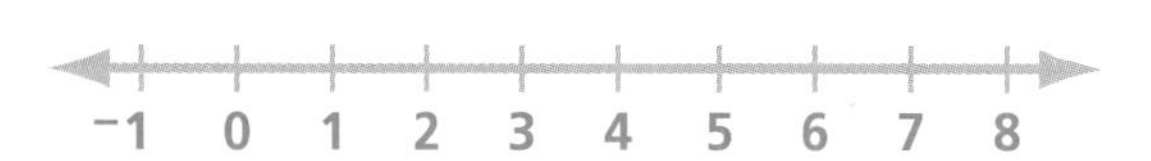

4. $5 < d < 7$

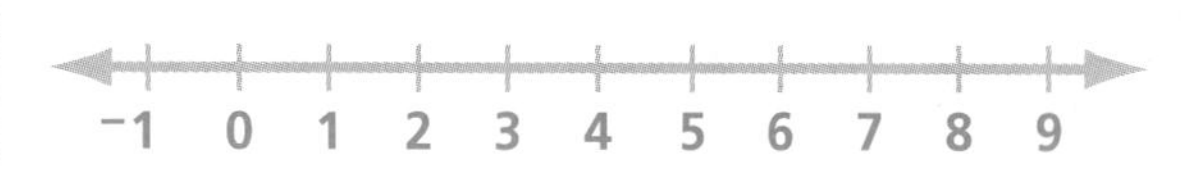

5. $^{-}2 \leq x \leq 7$

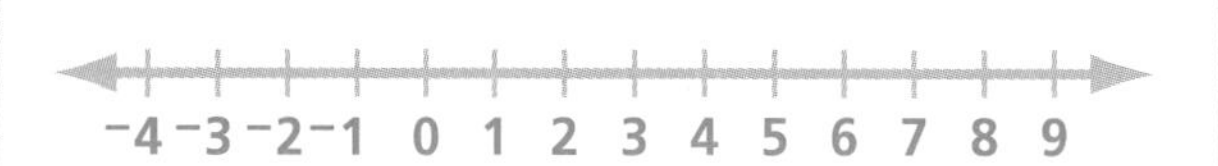

6. $n \leq {}^{-}1$

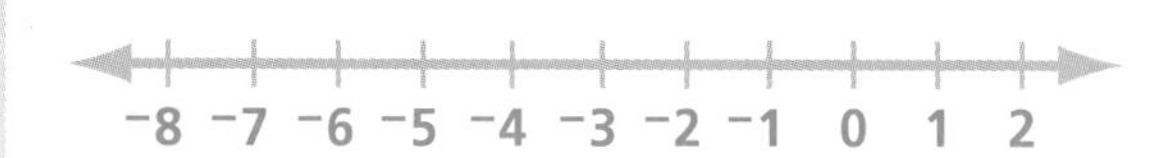

Solve each inequality. Graph the solution.

7. $y - 5 < {}^{-}2$

$y - 5 + 5 < {}^{-}2 + 5$

$y < 3$

8. $w - 1 > 8$

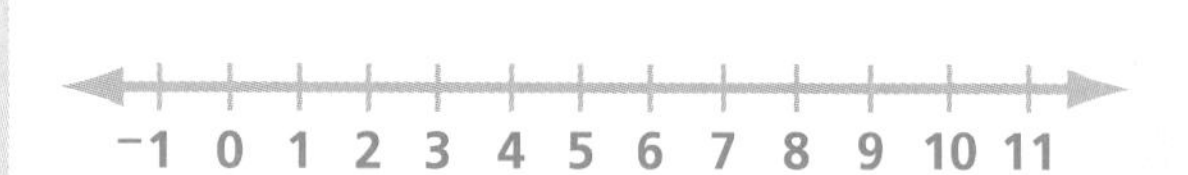

9. $1 \leq x - 4$

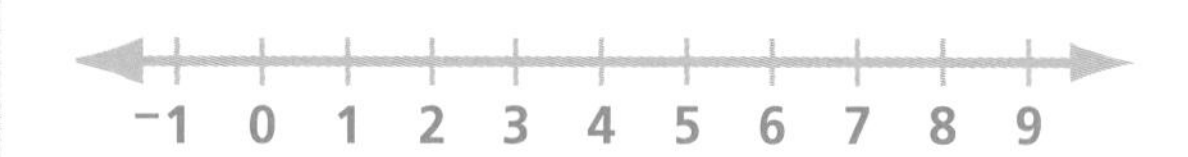

10. $x + 5 > 0$

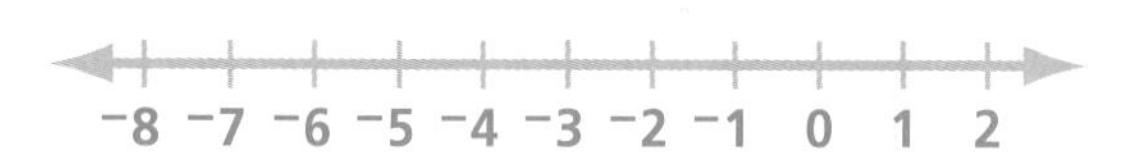

11. $3a - 3 \geq 2a + 1$

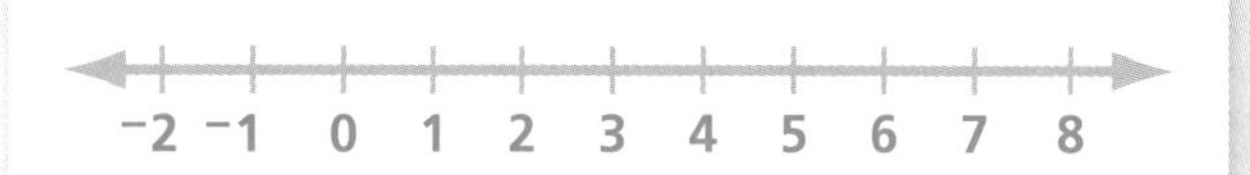

12. $6 + b \leq 5$

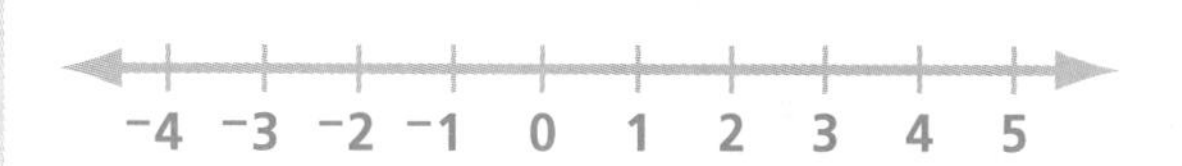

Describe the numbers modeled on the number line. Then write an inequality statement to describe the numbers.

____

____

Lesson 13 Multiplication and Division Inequalities

Multiplication and division inequalities are solved the same way as multiplication and division equations.

Solve each inequality.

Model 1

$3x \geq 9$

$\frac{3x}{\_\_\_\_} \geq \frac{9}{\_\_\_\_}$

$x \geq$ ______

$\frac{x}{4} < {}^{-}2$

$4 \cdot \frac{x}{4} < 4 \cdot {}^{-}2$

$x <$ ______

Multiplying or dividing both sides of the inequality by a negative number affects the inequality sign. Consider the following.

$2 < 4$

${}^{-}2 \cdot 2 \;\mathbf{?}\; {}^{-}2 \cdot 4$

${}^{-}4$ ______ ${}^{-}8$

Notice that the inequality sign changed. This will always happen when multiplying or dividing by a negative number to solve an inequality.

Solve each inequality.

Model 2

${}^{-}3x \geq 9$

$\frac{{}^{-}3x}{{}^{-}3} \leq \frac{9}{{}^{-}3}$

$x \leq$ ______

Think: Isolate the variable first.

Divide both sides by ${}^{-}3$; change the inequality sign.

$\frac{x}{{}^{-}4} < {}^{-}2$

${}^{-}4 \cdot \frac{x}{{}^{-}4}$ ______ ${}^{-}4 \cdot {}^{-}2$

x ______ ______

Think: What is the denominator?

Multiply both sides by ${}^{-}4$; change the inequality sign.

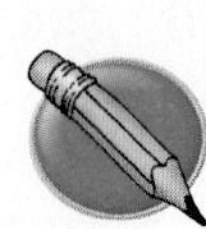

How is solving inequalities different from solving equations? How are they alike?

____

____

____

Practice

Solve each inequality. Show each step.

1. $7a > 49$

   $\frac{7a}{7} > \frac{49}{7}$

   $a > 7$

2. $3c < {}^{-}24$

3. $\frac{m}{7} < {}^{-}14$

4. $\frac{n}{6} \geq 2$

5. $5y \leq 75$

6. $\frac{r}{3} > {}^{-}18$

7. $9k < 0$

8. $4b > 1$

9. $12 < 3p$

Solve each inequality. Show each step.

10. ${}^{-}4n > {}^{-}40$

    $\frac{{}^{-}4n}{{}^{-}4} < \frac{{}^{-}40}{{}^{-}4}$

    $n < 10$

11. $\frac{d}{{}^{-}10} < {}^{-}10$

12. ${}^{-}3b > {}^{-}2$

13. $\frac{y}{{}^{-}3} \geq 13$

14. ${}^{-}9h \geq {}^{-}63$

15. $\frac{n}{{}^{-}5} \leq 3$

Describe how to check that the solution to an inequality is correct.

____

____

____

Lesson 14 Multi-Step Inequalities

Steps that are used to solve two-step equations are also used to solve two-step inequalities. Remember to change the inequality symbol when multiplying or dividing by a negative number.

Solve each inequality.

Model 1

$2x - 1 > 5$	**Think:** Combine the constants first.
$2x - 1 +$ ______ $> 5 +$ ______	Add 1 to both sides.
$2x >$ ______	Combine terms.
$x >$ ______	Divide both sides by 2.

$\frac{x+2}{5} < 3$	**Think:** Get rid of the denominator first.
$5 \cdot \frac{x+2}{5} < 5 \cdot$ ______	Multiply both sides by 5.
$x + 2 <$ ______	Subtract 2 from both sides.
$x <$ ______	

Model 2

$x + 3 \leq {}^{-}8x$	
$x + 3 -$ ______ $\leq {}^{-}8x -$ ______	Subtract 3 from both sides.
$x \leq {}^{-}8x -$ ______	Combine terms.
$x +$ ______ $\leq {}^{-}8x +$ ______ $-$ ______	Add $8x$ to both sides.
______ $\leq$ ______	Combine terms.
$x \leq$ ______	Divide both sides by 9.

Explain how to determine if $^{-}1$ is a solution to $3x - 2 > 4$.

____

____

____

Practice

Solve each inequality. Show each step.

1. $10 < 3 - 7n$

$10 - 3 < 3 - 3 - 7n$

$7 < {}^{-}7n$

$n < {}^{-}1$

2. $0 \geq 5 - c$

3. $30 \geq 6t - 18$

4. $\frac{y}{4} + 1 \geq \frac{3}{4}$

5. $\frac{d}{2} - 9 \leq 3$

6. $7p + 9 \leq {}^{-}5$

7. $12 + 3z > {}^{-}3$

8. $0 < \frac{x}{4} - 4$

9. ${}^{-}n + 5 \leq 15$

10. $2y + 1 > y$

11. ${}^{-}7k - 7k \leq {}^{-}7$

12. $w - \frac{1}{2} < \frac{3w}{2}$

13. ${}^{-}r + 14 \leq r + 12$

14. $3b - 2 < 2(4 - b)$

15. $\frac{3n}{2} - 9 < 3 + n$

Explain the steps necessary to solve $5x - 4 < 4(8 - x)$. Then solve the inequality.

____

____

Strength Builder

▶ Fun with Codes

Pssst . . . Want to Know a Secret?

A cryptologist is a person who creates and breaks secret codes. Try your hand at cryptology by decoding the message below.

In this coded message, each number stands for a different letter of the alphabet. Solve the equations to find the number that matches each letter. Then substitute the letters for the numbers to break the code and reveal the message.

A $^-2a = 6a + 2$

$^-2a - 6a = 6a - 6a + 2$

$\frac{^-8a}{^-8} = \frac{2}{^-8}$

$a = \frac{^-1}{4}$

C $2c + 5 = 1$

D $6(d - 3) = {}^-3d$

F $3f - 5 = 15f + 10$

G $\frac{^-8g}{7} = 16$

H $\frac{2h}{3} + \frac{1}{3} = 15$

K $k - 3 = 12$

M $^-6 = {}^-2m - 8$

N $n - 9n = 64$

R $5r = 100$

S $3s + 10 = 5s$

W $\frac{6 - 2w}{4} = {}^-3$

Break the Code

Complete the table below using solutions from the equations.

A	B	C	D	E	F	G	H	I
______	$\frac{1}{2}$	______	______	3	______	______	______	$^{-}9$
J	**K**	**L**	**M**	**N**	**O**	**P**	**Q**	**R**
12	______	$^{-}6$	______	______	7	$\frac{1}{3}$	$\frac{3}{4}$	______
S	**T**	**U**	**V**	**W**	**X**	**Y**	**Z**	
______	$^{-}11$	8	6	______	19	100	$\frac{1}{5}$	

Use the numbers from the chart to decode the message.

Message:

$^{-}11$ 22 3 $^{-}11$ 20 3 $^{-}\frac{1}{4}$ 5 8 20 3

$^{-}9$ 5 $\frac{1}{2}$ 8 20 $^{-}9$ 3 2 $^{-}11$ 3 $^{-}8$

$\frac{1}{3}$ $^{-}\frac{1}{4}$ $^{-}2$ 3 5   9 3 5 $^{-}11$

7 $^{-}\frac{5}{4}$ $^{-}11$ 22 3 $\frac{1}{3}$ 3 $^{-}2$ $^{-}\frac{1}{4}$ $^{-}8$

$^{-}11$ 20 3 3 .

Unit

2 Review

Complete the equation to show each property.

1. Associative Property

$a \cdot (b \cdot c) =$ ______________

2. Commutative Property

$a + b =$ ______________

3. Identity Property of Addition

$a + 0 =$ ______________

4. Distributive Property

$a(b + c) =$ ______________

5. Inverse Property of Addition

$a +$ ______________ $= 0$

6. Identity Property of Multiplication

$a \cdot$ ______________ $= a$

Simplify each expression.

7. $8 - 14 \cdot 2 \div 7 + 10$

8. $\frac{12x - 17x}{5}$

9. $4(2b - 7) - 8b$

Solve each equation.

10. $^{-}3n + 6 = {^{-}18}$

11. $^{-}10p + 6 = 4p - 22$

12. $\frac{a}{2} - \frac{3}{4} = \frac{1}{8}$

Solve each inequality.

13. $x + 3 \leq 1$

14. $\frac{m}{^{-}3} > {^{-}2}$

15. $2y - 7 \geq {^{-}1}$

Write an equation to model each situation. Solve the equation.

16. Kara scored 9, 13, and 6 points in her last three games. How many points must she score in her fourth game to bring her average points per game to 10?

17. Juan puts the same amount into savings every week. In 12 weeks he has saved $240. How much does he save every week?

Unit 3

Graphs and Functions

What's inside a computer?

While you are downloading a file from the Internet, the data may be traveling at 500 bits per second.

For a circuit board to work correctly, the input and output from all the parts on the board must work together. People who design boards must plan the flow of the data logically.

What are some ways mathematics is used to describe the working of a computer?

Linear Equations

The coordinate system is based on the intersection of a horizontal line called the **x-axis** and a vertical line called the **y-axis**. The point of intersection is called the **origin**. The intersecting axes divide the plane into four **quadrants**.

Each point has a unique set of coordinates that describes its location.

Model 1

Look at point $A(3, 2)$. The 3 represents 3 units to the right from the origin on the ________, and the 2 represents 2 units above the origin. There is only one point that satisfies both of these conditions.

Write the coordinates of the following points.

Point B ______

Point C ______

Point D ______

Origin ______

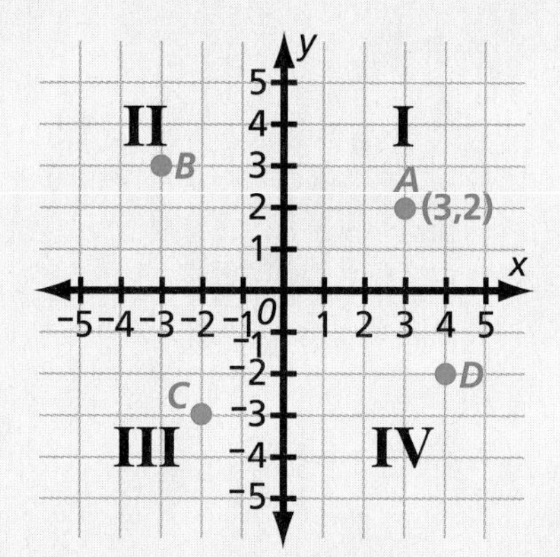

Model 2

The points (1, 2), (0, 0), and ($^-1$,$^-2$) are on the graph of $y = 2x$.

Graph these points on the coordinate plane.

Notice that the three points lie on the same line. This means that $y = 2x$ is a **linear equation**—an equation whose graph is a line.

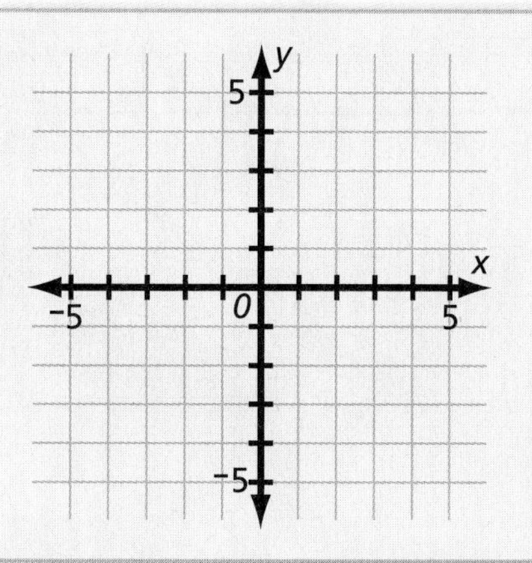

Explain why (4, 5) and (5, 4) are not the same point on the coordinate plane.

____

____

Practice

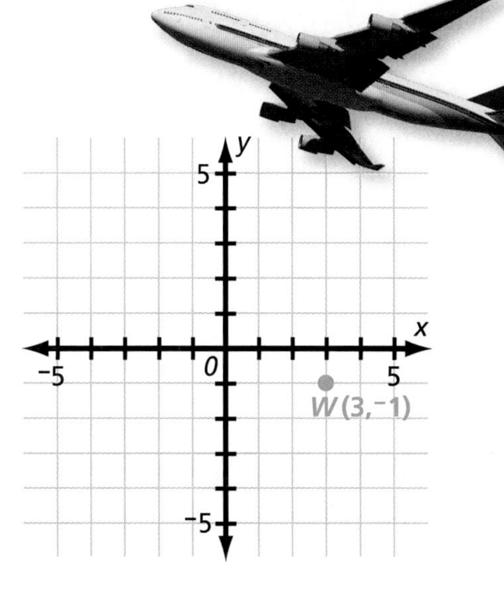

Plot each point and label the coordinates.

1. $W(3, ^-1)$

2. $X(^-4, 0)$

3. $Y(0, ^-2)$

4. $Z(^-3, ^-5)$

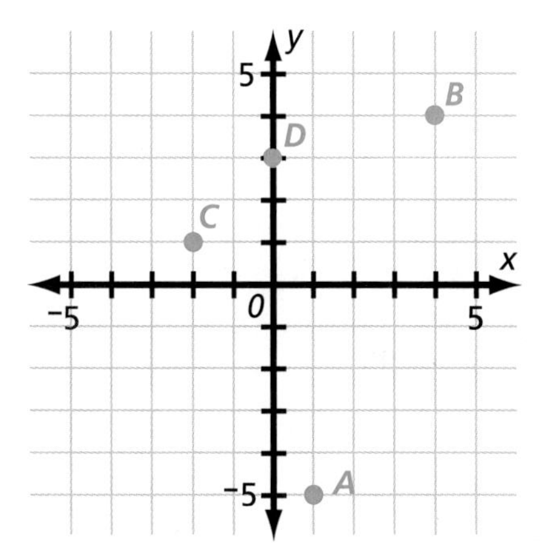

Identify the coordinates of each point.

5. Point A = **(1, ⁻5)**

6. Point B =

7. Point C =

8. Point D =

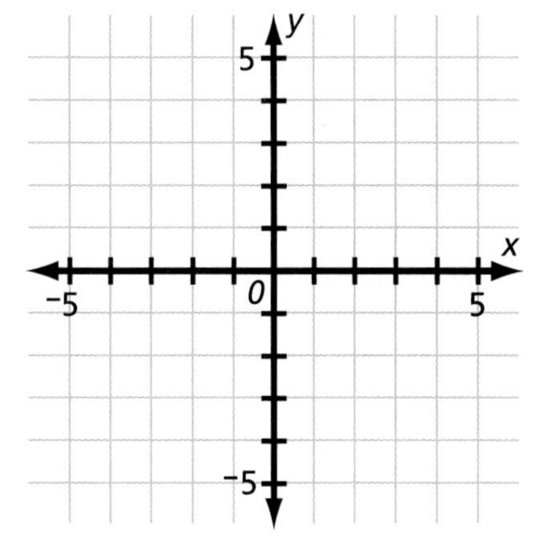

Plot each set of points. If the graph represents a linear equation write *linear* in the blank. Write *nonlinear* if it does not represent a linear equation.

9. $(0, ^-3)$, $(^-2, ^-5)$, $(1, ^-2)$ linear

10. $(1, 2)$, $(0, 1)$, $(2, 5)$ ___________

11. $(^-2, ^-5)$, $(4, 4)$, $(2, 1)$ ___________

The points (0, 3), (2, 7), and (4, 11) are on the same line. Explain how to find the coordinates of another point on the same line.

____

____

Lesson 2

Linear Equations and Graphs

Some problems can be modeled by writing and graphing a linear equation.

Use a dot to indicate multiplication in the equation.

Model 1

Beth works in a factory making candles. At the beginning of each day she spends 20 minutes setting up her workstation. Beth takes 2 minutes to make each candle. **Write an equation that describes the amount of time it takes Beth to produce a specific number of candles.**

Total time	=	**Time for one candle**	·	**Number of candles**	+	**Set-up time**
y	=	______	·	x	+	______

Model 2

Complete the table and draw a graph to model Beth's work output.

Graph these points to see the graph of the linear equation.

x	$y = 2x + 20$	y	(x, y)
0	$y = 2 \cdot$ ___ $+ 20$		
5	$y = 2 \cdot$ ___ $+ 20$		
10	$y = 2 \cdot$ ___ $+ 20$		

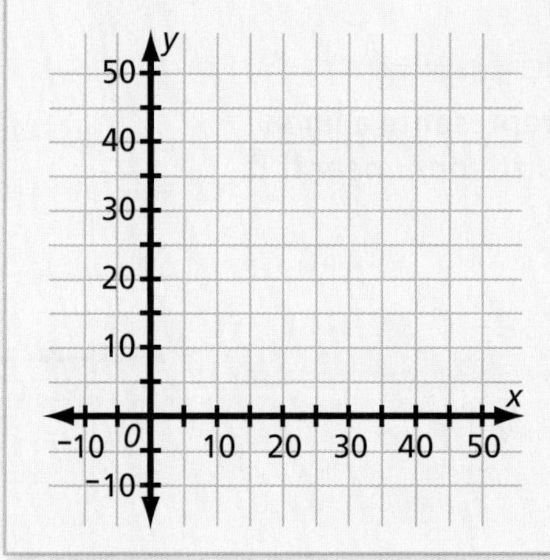

Use the line to determine how many candles Beth can make in the first 38 minutes of the day.

When $y = 38$, $x =$ ______.

In 38 minutes, Beth can make ______ candles.

Explain why only points in Quadrant I make sense for this problem.

____

____

Practice

Write an equation that could be used to answer the question.

1. On Friday the City Gift Shop donates one-tenth of the day's total sales plus $200 to charity. What is the total donation?

   $y = \frac{1}{10}x + 200$

2. Gina earns $100 a week and 4% commission on the total price of any merchandise she sells. How much did Gina earn this week?

3. The Pizza Shop charged $5 for each pizza for a party. What was the total cost?

4. A collector started with 20 books and added 1 book a month to his collection. How many books does he have now?

Complete each table of values. Then graph the equation to complete each statement.

5. $y = {}^{-}x$

x	$y = {}^{-}x$	y	(x, y)
$^{-}2$	$y = {}^{-}({}^{-}2)$	2	$({}^{-}2, 2)$
0	$y =$		
2	$y =$		

When $y = {}^{-}4$, the value of x is ______.

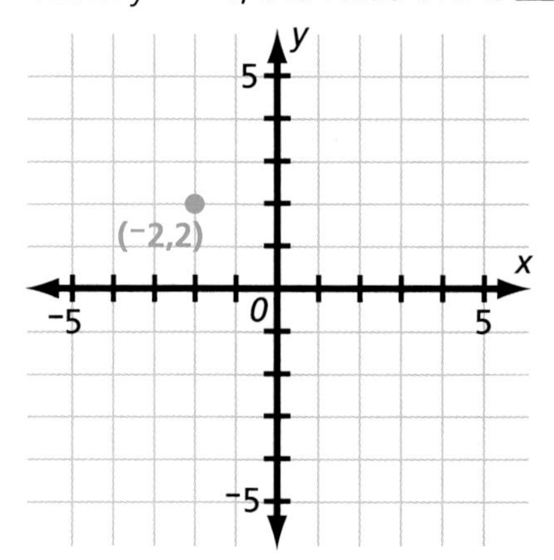

6. $y = x - 2$

x	$y = x - 2$	y	(x, y)
$^{-}3$	$y =$ ___ -2		
0	$y =$ ___ -2		
3	$y =$ ___ -2		

When $y = {}^{-}1$, the value of x is ______.

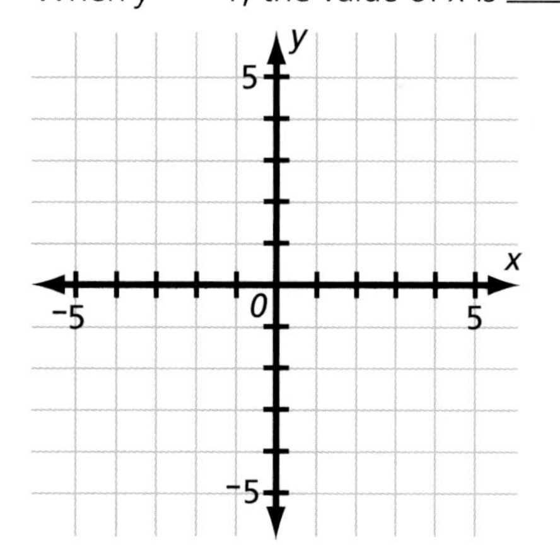

Show that $y = x^2 + 1$ is not a linear equation by creating a table of values and graphing the points on your own graph paper.

Slope of Lines

The **slope** of a line describes the line's steepness and direction.

The slope is the ratio of the **rise** (vertical change) to the **run** (horizontal change) between points on the line.

$\text{slope} = \frac{\text{rise}}{\text{run}}$

Use the graph of $y = 3x$ to find the slope.

Model 1

Mark two points. Compare rise to run.

$\text{slope} = \frac{\text{rise}}{\text{run}} = \frac{3}{1} =$ ______

The line $y = 3x$

has slope = ______.

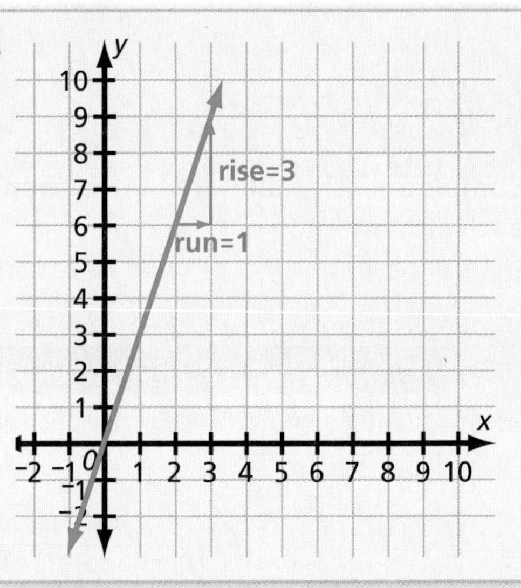

Use the formula slope $= \frac{\text{change in } y \text{ (rise)}}{\text{change in } x \text{ (run)}} = \frac{y_2 - y_1}{x_2 - x_1}$.

Model 2

Find the slope of the line containing the points (4, 2) and (6, 3). Let (4, 2) = (x_2, y_2) and let (6, 3) = (x_1, y_1).

$\text{slope} = \frac{\text{change in } y}{\text{change in } x} = \frac{y_2 - y_1}{x_2 - x_1} = \frac{2 - 3}{4 - 6} =$ ______ $=$ ______

The line through (4, 2) and (6, 3) has a slope of $\frac{1}{2}$.

Explain how to find the slope of the line containing the points (2, ⁻3) and (5, ⁻3).

____

____

Practice

Use the graph to find the slope of various lines.

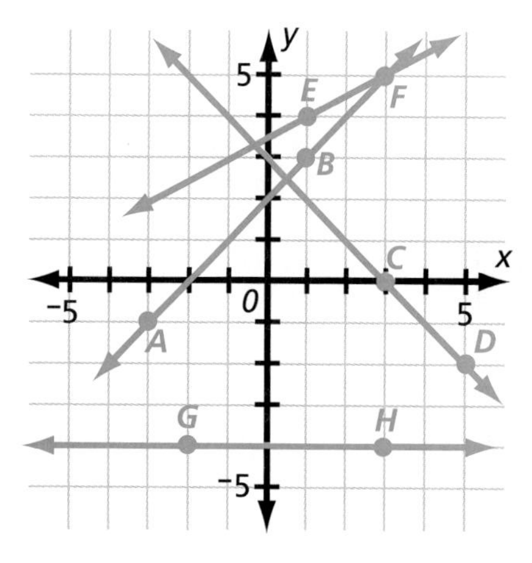

1. slope of line $AB = \frac{4}{4} = 1$

2. slope of line $CD =$

3. slope of line $EF =$

4. slope of line $GH =$

Determine the slope of the line passing through the given points.

Use the formula slope $= \frac{y_2 - y_1}{x_2 - x_1}$.

5. $K(3, 4)$, $L(0, 0)$

   $\frac{4-0}{3-0} = \frac{4}{3}$

   slope: $\frac{4}{3}$

6. $M(4, {}^-1)$, $N(1, 2)$

   slope:

7. $P(4, {}^-4)$, $Q(7, {}^-2)$

   slope:

8. $R(2, {}^-9)$, $S(0, 2)$

   slope:

9. $T(5, {}^-1)$, $U(7, 3)$

   slope:

10. $W(8, 0)$, $V(6, {}^-2)$

    slope:

If two different lines have the same slope, what must be true about the lines?

___

___

Lesson 4 Slopes and Intercepts

An **intercept** is the point where the graph of an equation crosses an axis. The point where the graph crosses the x-axis is called the x-intercept. The point where the graph crosses the y-axis is called the y-intercept.

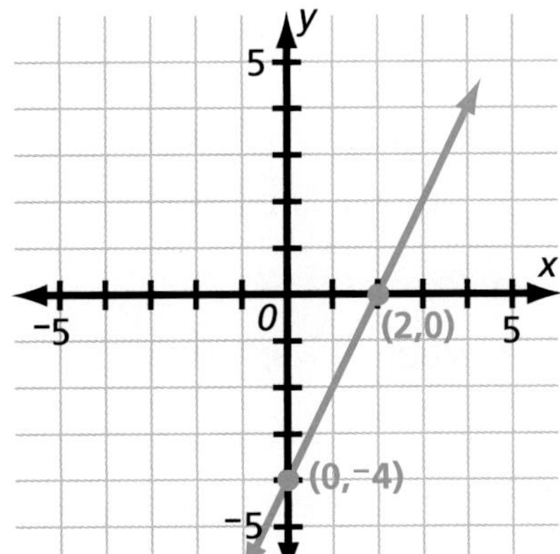

The x-intercept is ______ and the

y-intercept is ______.

The form $y = mx + b$ is called **slope-intercept form**. When an equation is written this way, m equals the slope of the line and b equals the y-intercept of the line.

Find each slope (m) and y-intercept (b), since $y = mx + b$.

Model 1

The line given by $y = 2x - 4$ has a slope (m) of ______ and a y-intercept (b) of ______.

The line given by $y = {}^{-}5x + 6$ has a slope (m) of ______ and a y-intercept (b) of ______.

Use the slope and the y-intercept to write the equation of a line.

Model 2

You can write the equation of a line when you know the slope and the y-intercept.

$$m = \frac{{}^{-}1 - ({}^{-}2)}{0 - 1} = \frac{1}{{}^{-}1} = \text{______},$$

y-intercept = ______

The equation of the line is

______________________.

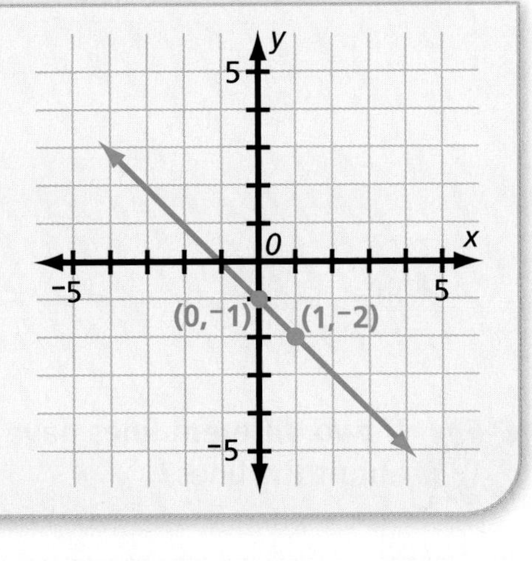

If several coordinate pairs have the same y-value, will the graph of the points be a horizontal or vertical line? Explain.

____

____

Practice

Identify the *x*- and *y*-intercept of each line.

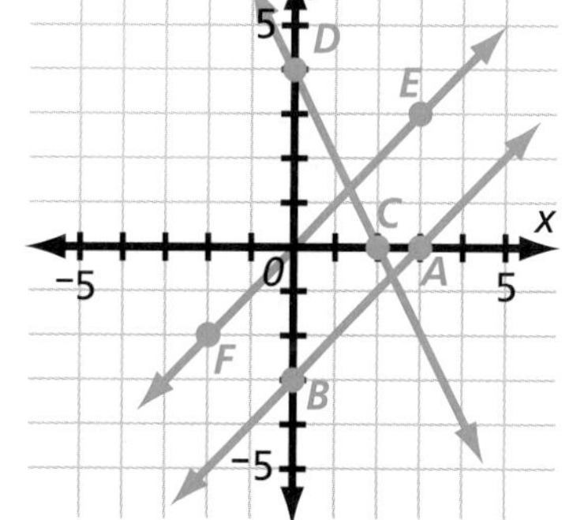

1. Line *AB*

x-intercept: 3 *y*-intercept: $^-3$

2. Line *CD*

x-intercept: ______ *y*-intercept: ______

3. Line *EF*

x-intercept: ______ *y*-intercept: ______

Identify the slope and the *y*-intercept of each line.

4. $y = \frac{7}{3}x + 10$

$m = \frac{7}{3}$

$b = 10$

5. $y = 5x + 9$

$m =$

$b =$

6. $y = {}^-8x - 2$

$m =$

$b =$

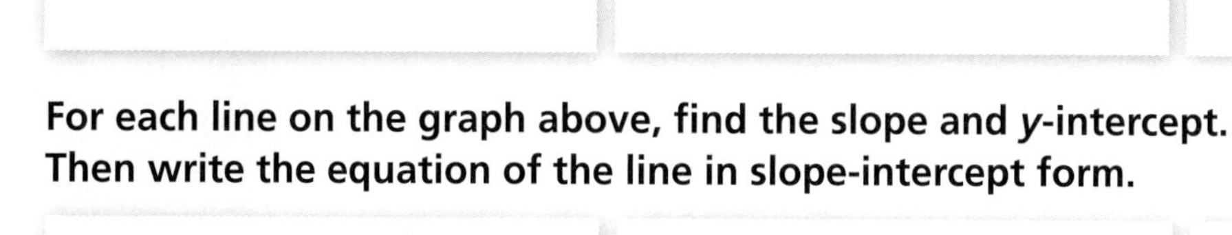

7. $y = -\frac{4}{7}x + \frac{4}{7}$

$m =$

$b =$

8. $y = {}^-3x - 4$

$m =$

$b =$

9. $y = {}^-x$

$m =$

$b =$

For each line on the graph above, find the slope and *y*-intercept. Then write the equation of the line in slope-intercept form.

10. Line *AB*

$m = 1$

$b = {}^-3$

equation: $y = x - 3$

11. Line *CD*

$m =$

$b =$

equation:

12. Line *EF*

$m =$

$b =$

equation:

A line contains the points (2, 6) and ($^-5$, 6). Explain how to find the slope and the *y*-intercept of the line. Then write the equation of the line in slope-intercept form.

____

____

Lesson 5

Direct Variation

If the ratio of two variables is a nonzero constant, then the two variables are said to have **direct variation**.

Example: Carlo drives 45 miles on the highway and uses 3 gallons of gasoline. The number of miles Carlo drives (y) varies directly with the amount of gasoline he uses (x). The variables vary directly because as the value of x increases, the value of y also increases.

Model 1

When two variables vary directly, the ratio of the variables equals a constant (k). This is called the **constant of variation**.

General form: $y = kx$

Find the constant of variation where $y = 45$ and $x = 3$.

$y = kx$

$45 = k \cdot 3$

$\frac{45}{3} = k$

$k =$ ______

Model the situation with an equation and find the constant of variation.

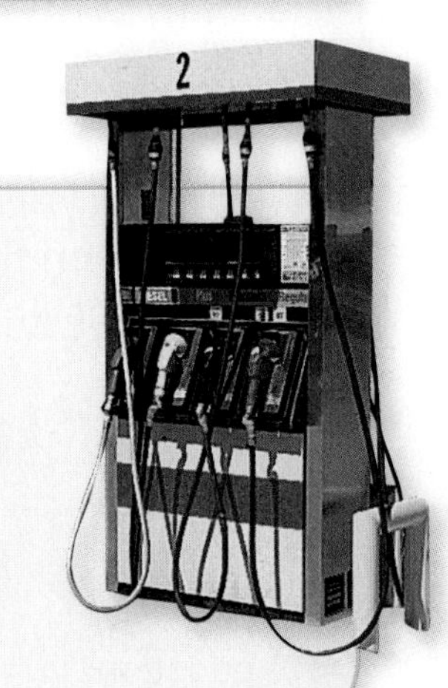

Model 2

Dale earns an hourly wage. Last week he worked 30 hours and earned $240.

$y = kx$

$240 = k \cdot 30$

$\frac{240}{30} = k$

$k =$ ______

What is the relationship between the general form for direct variation, $y = kx$ and the linear equation $y = mx + b$?

____

____

Practice

Find the constant of variation for each situation.

1. $y = 35, x = 5$

   $k = 7$

2. $y = 100, x = 2$

   $k =$

3. $y = {}^{-}60, x = 5$

   $k =$

4. $y = 2, x = 4$

   $k =$

5. $y = {}^{-}55, x = {}^{-}11$

   $k =$

6. $y = 40, x = 40$

   $k =$

Model the situation with an equation and find the constant of variation.

7. Apples are priced by the pound. A 5-pound bag of apples costs \$2.00.

   \$2.00 = $k \cdot 5$

   k = \$0.40

8. Party favors are sold by the bag. Nine bags of favors cost \$4.50.

9. Ribbon is needed for each costume. Forty yards of ribbon are needed for 10 costumes.

10. Each student is given a set of papers. Nine hundred papers are given to 150 students.

11. Muffins are sold in small containers. Thirty-six muffins fill 9 containers.

12. Teams were organized for the relay race. Eighty students made up 10 teams.

Describe an everyday situation in which two variables vary directly. Then write an equation to model the situation.

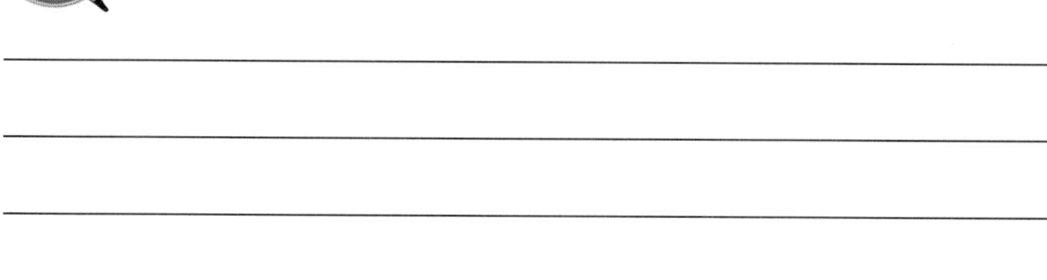

Lesson 6

Tables of Solution Sets

The group of all the solutions to an equation is called the **solution set**. Using a table is a good way to organize a solution set.

Complete the table for the equation $y = x - 1$.

Model 1

x	$y = x - 1$	y	(x, y)
0	$y =$ ______ $- 1$	$^-1$	
1	$y =$ ______ $- 1$		
$^-1$	$y =$ ______ $- 1$		
2	$y =$ ______ $- 1$		

Each (x, y) pair of numbers is a solution to the equation. Each pair is a member of the solution set.

Verify that an ordered pair is part of the solution set of an equation by substituting the values into the equation.

Determine whether each pair is a solution of $y = 3x - 5$.

Model 2

(0, 25)

$y = 3x - 5$

$25 \neq 3 \cdot$ ______ $- 5$

______ $\neq$ ______

(0, 25) is not a solution to the equation.

(2, 1)

$y = 3x - 5$

______ $= 3 \cdot$ ______ $- 5$

______ $=$ ______ $- 5$

______ $=$ ______

(2, 1) is a solution to the equation.

What would a table of solutions for $x = 5$ look like?

____

____

____

Practice

Complete the table for each equation.

1.

x	$y = 4x - 10$	y	(x, y)
$^{-}4$	$y = 4 \cdot$ ___$^{-}4$___ $- 10$	$^{-}26$	$(^{-}4, {}^{-}26)$
$^{-}2$	$y = 4 \cdot$ ______ $- 10$		
0	$y = 4 \cdot$ ______ $- 10$		
2	$y = 4 \cdot$ ______ $- 10$		

2.

x	$y = \frac{1}{5}x$	y	(x, y)
$^{-}5$	$y = \frac{1}{5} \cdot$ ______		
0	$y = \frac{1}{5} \cdot$ ______		
5	$y = \frac{1}{5} \cdot$ ______		
10	$y = \frac{1}{5} \cdot$ ______		

Determine whether each pair is a solution to the equation.

3. $(4, 1)$

$y = 4x - 1$

$4 \cdot 4 - 1 = 15;$

$15 \neq 1$; not a solution

4. $(3, 3)$

$y = {}^{-}x$

5. $(^{-}2, 1)$

$y = x + 3$

6. $(6, 7)$

$y = \frac{2}{3}x + 3$

7. $(^{-}2, 4)$

$y = {}^{-}x - 6$

8. $(12, {}^{-}10)$

$y = {}^{-}x + 2$

What would be true about the graphs of two equations that had the same solution set?

____

____

Solution Sets and Graphs

The solution set of an equation can be graphed on the coordinate plane. A table of values is helpful when making the graph.

Graph the solution set of $y = 3x + 1$.

Model 1

Graph the (x, y) pairs from the table. Draw a line through the points.

x	y	(x, y)
$^-2$		
$^-1$		
1		

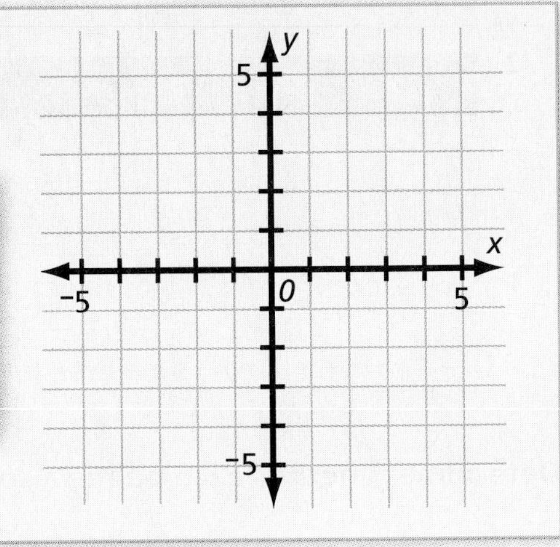

Substitute the values for x and y into the equation.

Model 2

Is the point (3, 10) on the graph of $y = 3x + 1$?

$10 = 3 \cdot$ ______ $+ 1$

$10 =$ ______ $+ 1$

$10 =$ ______ (3, 10) is a solution to the equation $y = 3x + 1$.

Find the y-intercept on the graph of $y = 3x + 1$.

Model 3

When the line crosses the y-axis, the value of the x-coordinate is 0.

$y = 3 \cdot$ ______ $+ 1$

$y =$ ______ The y-intercept is ______.

Describe how to find the x-intercept from the equation $y = 3x + 1$.

____

____

Practice

Complete the table. Then draw the graph of the equation.

1. $y = {}^{-}2x - 1$

x	y	(x, y)
$^{-}2$	3	$(^{-}2, 3)$
0		
1		
2		

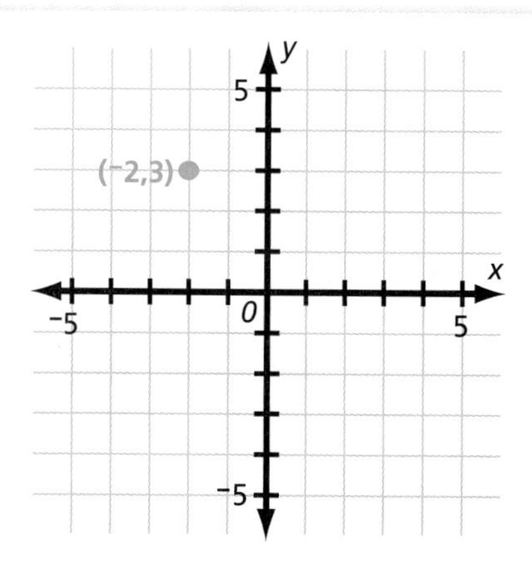

Determine whether each point is part of the solution set of $y = {}^{-}2x - 1$.

2. $(^{-}3, 8)$

$^{-}2 \cdot {}^{-}3 - 1 = 6 - 1 = 5$

$5 \neq 8$; not a solution

3. $(^{-}1, 1)$

4. $(\frac{1}{2}, {}^{-}2)$

5. $(4, {}^{-}9)$

6. $(\frac{11}{2}, {}^{-}12)$

7. $(^{-}8, 15)$

Complete the table below.

	x	$y = 2x + 4$	(x, y)	intercept
8.	0			y-intercept
9.		0		

How does graphing the line representing an equation help find other solutions to the equation?

____

____

____

____

Lesson 8

Graphs of Inequalities with Two Variables

Inequalities can be written with two variables.

Examples: $y > x + 4$
$y \geq x - 3$
$2y + 3x \leq 7$

Inequality Symbols	
$<$	less than
$>$	greater than
$\leq$	less than or equal to
$\geq$	greater than or equal to

Notice that these look like equations for lines, except that the equal signs are replaced by inequality symbols. These are called **linear inequalities**.

Model 1

At right is a graph of $y < x + 4$. The line $y = x + 4$ forms what is called the **boundary line**. Because this inequality involves $<$, and not $\leq$, the line is dashed. The points on the line are *not* included in the solution set.

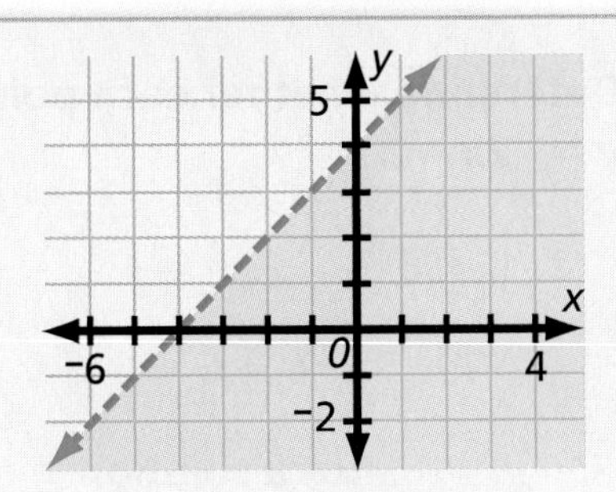

Graph $y \geq 3x - 2$.

Model 2

Graph the equation $y = 3x - 2$ first to form the boundary line. The line will be solid, because the $\geq$ indicates that the line is included in the solution set.

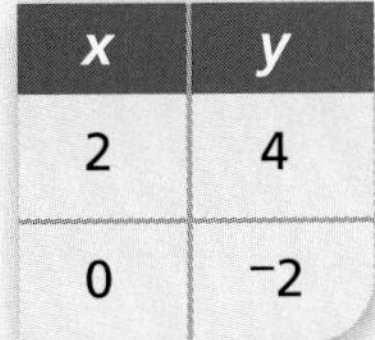

x	y
2	4
0	−2

The second step is to decide which side of the line to shade. Choose a point on one side or the other of the line to test. Is (0, 0) in the solution set?

$y \geq 3x - 2$ ______ $\geq$ ______

The statement is true, so (0, 0) is part of the solution set.

Shade the part of the graph containing (0, 0).

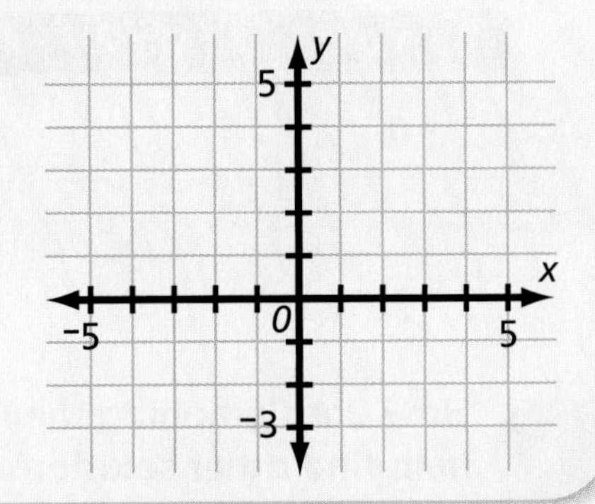

What does a solid boundary line indicate?

____

____

Practice

Graph each inequality. Write *dashed* or *solid*. Circle *is* or *is not*.

1. $y > x + 4$

x	y
1	5
0	
$^-1$	

The boundary line should be ______________.

The point (0, 0) (is, is not) part of the solution set.

2. $y \geq {}^-x - 3$

x	y
1	
0	
$^-1$	

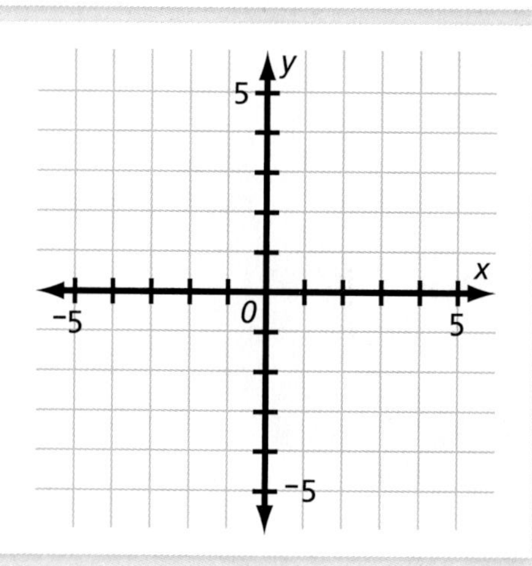

The boundary line should be ______________.

The point (0, 0) (is, is not) part of the solution set.

3. $y \geq \frac{1}{2}x + 1$

x	y
2	
0	
$^-2$	

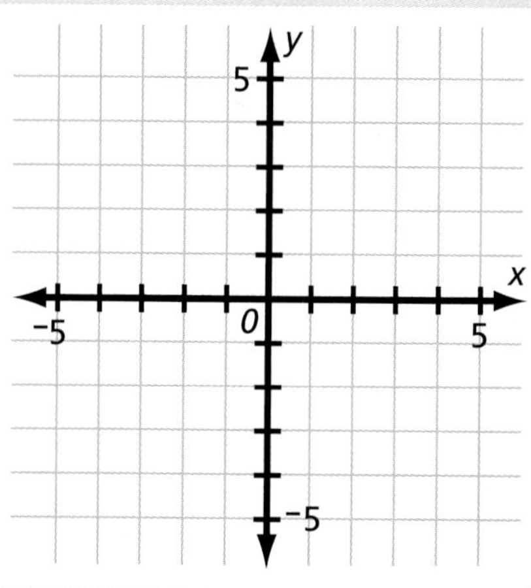

The boundary line should be __________.

The point (0, 0) (is, is not) part of the solution set.

Explain what the graph of $y \leq 3$ would look like.

____

____

Lesson 9 Systems of Equations

A **system of equations** is two or more equations that use two or more variables. The solution to a system of linear equations is the ordered pair for the variables that makes the equations true.

Solve the system of equations $\begin{matrix} y = {}^{-}x + 3 \\ y = x + 1 \end{matrix}$ **by graphing.**

Model 1

The point where the lines intersect is the solution to the system of equations.

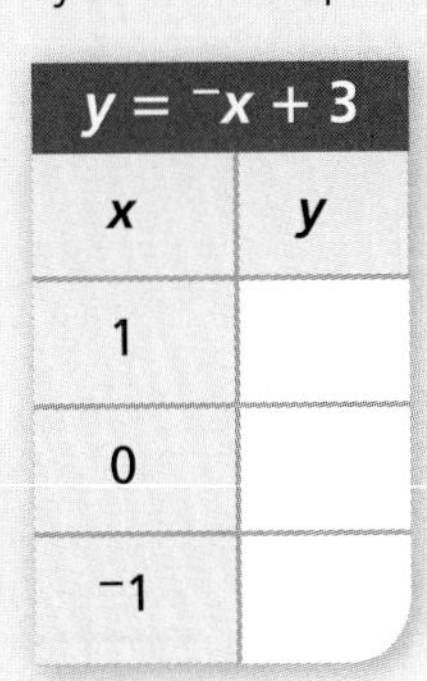

$y = {}^{-}x + 3$	
x	y
1	
0	
$^{-}1$	

$y = x + 1$	
x	y
1	
0	
$^{-}1$	

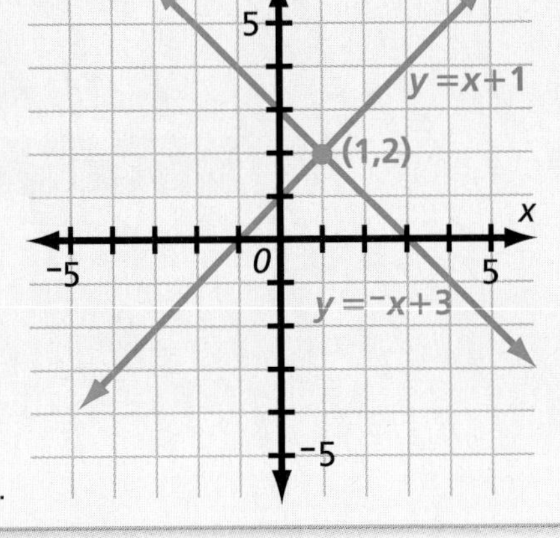

The solution to this system is ______.

Solve the system of equations $\begin{matrix} y = 2x + 3 \\ y = 2x - 1 \end{matrix}$ **by graphing.**

Model 2

$y = 2x + 3$	
x	y
1	
0	
$^{-}1$	

$y = 2x - 1$	
x	y
1	
0	
$^{-}1$	

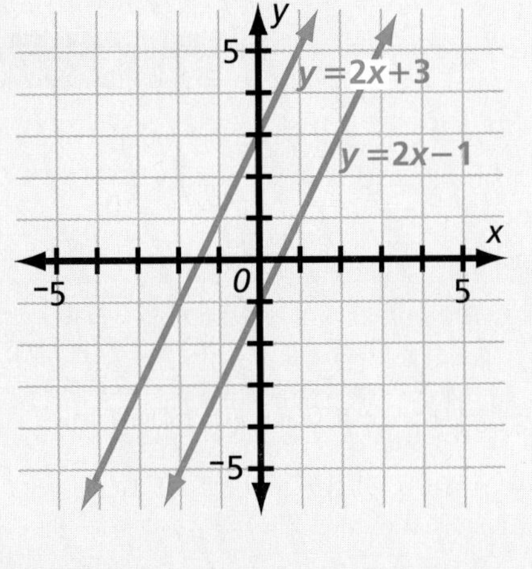

The lines do not intersect; they are ____________.

So there is ______ solution to the system of equations.

Describe the solution to the system of equations $y = 2x + 4$ and $3y = 6x + 12$. What is unique about the system?

____

____

Practice

Solve each system of equations by graphing. State the solution.

1. $y = x + 2$

$y = \frac{-1}{3}x - 2$

solution: **$(-3, -1)$**

$y = x + 2$	
x	y
0	2
-2	0
1	3

$y = \frac{-1}{3}x - 2$	
x	y
0	
3	
-3	

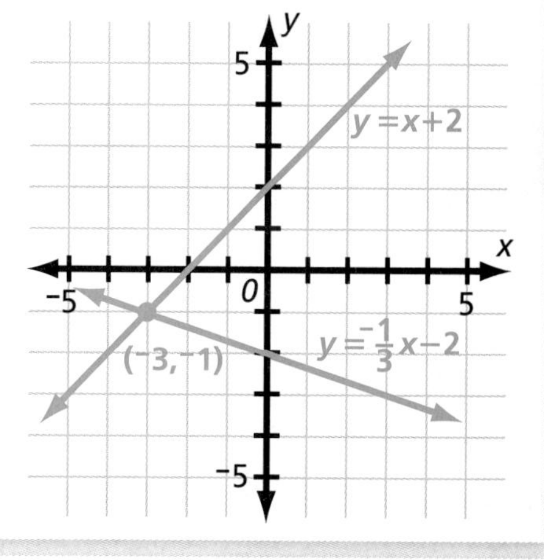

2. $y = -x + 2$

$y = 2x - 1$

solution:

$y = -x + 2$	
x	y
0	
2	
1	

$y = 2x - 1$	
x	y
2	
0	
-1	

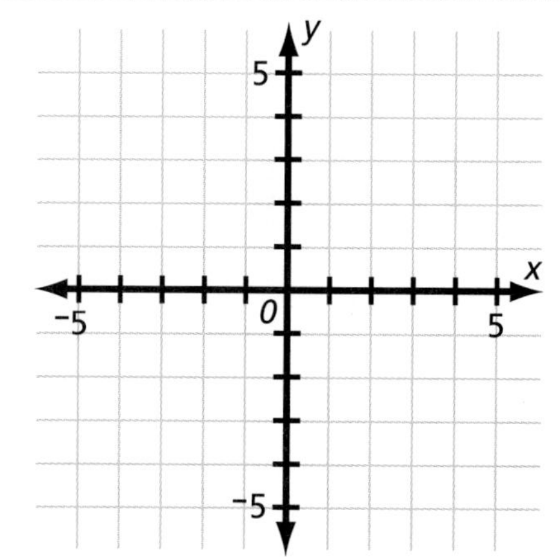

3. $y = x + 1$

$y = -x + 5$

solution:

$y = x + 1$	
x	y
0	
-1	
3	

$y = -x + 5$	
x	y
0	
5	
3	

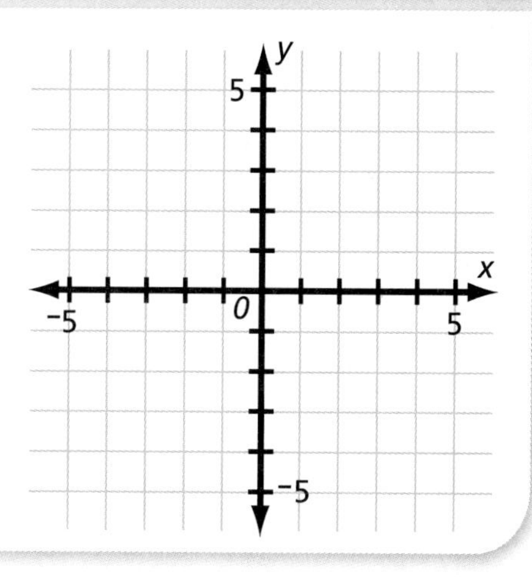

Explain how to determine if $(-1, 3)$ is the solution to the system of equations $\begin{matrix} y = -x + 2 \\ y = 2x + 5 \end{matrix}$.

Lesson 10

Linear Functions

A number rule that relates an input value, such as x, to a single output value, such as y, is called a **function**. The equation $y = x + 3$ states that for every x chosen as an input, adding 3 to x will give the output. The equation can also be called a *function of x*.

The phrase *function of x* describes the relationship between the input and the output of a number rule.

Replace y with $f(x)$, read *f of x*. Rewrite $y = x + 3$ as $f(x) = x + 3$. For every input value of x, the function generates one output value equal to $x + 3$. If $x = 5$, $f(5) = 5 + 3 = 8$. This is called **evaluating** a function.

Evaluate the function $f(x) = 3x - 2$ for each of the following values.

Model 1

$x = 4$	$x = {}^{-}\frac{2}{3}$
$f(x) = 3x - 2$	$f(x) = 3x - 2$
$f(4) = 3 \cdot 4 - 2$	$f(-\frac{2}{3}) = 3 \cdot {}^{-}\frac{2}{3} - 2$
$f(4) =$ ______ $- 2$	$f(-\frac{2}{3}) =$ ______ $- 2$
$f(4) =$ ______	$f(-\frac{2}{3}) =$ ______

The **domain** of a function is the set of all the input values that can be used for a function. The **range** of a function is the set of all the corresponding output values the function generates.

Model 2

Find the domain and range of $f(x) = {}^{-}x + 3$.

You can use any real number for x.

domain: ____________________

The output of the function will also be all real numbers.

range: ____________________

Explain what the domain and range of $f(x) = x^2$ would be.

____

____

Practice

Evaluate when $f(x) = \frac{3}{2}x - 2$.

1. $f(0) =$

   $f(0) = \frac{3}{2} \cdot 0 - 2$

   $= 0 - 2 = {}^{-}2$

2. $f(^{-}6) =$

3. $f(\frac{1}{3}) =$

4. $f(4) =$

5. $f(2) =$

6. $f(\frac{2}{3}) =$

Evaluate when $f(x) = 2x - 3$.

7. $f(0) =$

   $f(0) = 2 \cdot 0 - 3$

   $= 0 - 3 = {}^{-}3$

8. $f(^{-}6) =$

9. $f(\frac{1}{2}) =$

10. $f(4) =$

11. $f(2) =$

12. $f(^{-}3) =$

Find the domain and range.

13. $f(x) = x + 3$

    domain: all real numbers

    range: all real numbers

14. $f(x) = 6x$

15. $f(x) = x^2$

Describe the graph of a function that has a domain of all real numbers and a range of only the number 5.

____

____

____

Lesson 11

Data and Functions

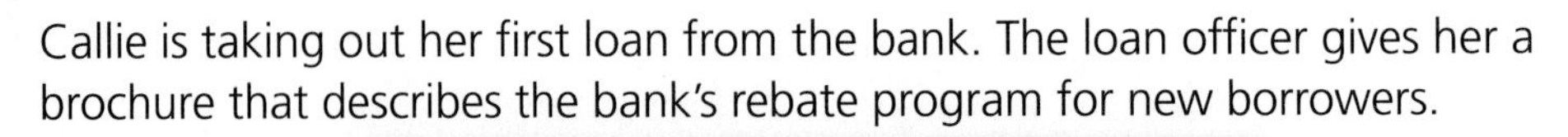

Callie is taking out her first loan from the bank. The loan officer gives her a brochure that describes the bank's rebate program for new borrowers.

If you borrow	Your rebate will be
\$500	\$ 5
\$1,000	\$10
\$1,500	\$15
\$2,000	\$20

The table shows a pattern. This pattern can be written as an expression, and then as a function using $f(x)$.

Model 1

Write the pattern as a function.

$500 \div$ ____ $= 5$	$1000 \div$ ____ $= 10$	$1500 \div$ ____ $= 15$	$2000 \div$ ____ $= 20$

Each value borrowed, x, is divided by ______. This can be written as

the expression $x \div 100$. Then $f(x) =$ __________.

Model 2

If Callie borrows \$3,750 what will her rebate be?

Evaluate *f*(3750).

$f(x) = x \div 100$

$f(3750) =$ ______ $\div 100$

$f(3750) =$ ______

Callie's rebate will be \$37.50

Model 3

Find the relationship between *x* and *f*(*x*) in the table. Then write the relationship as a function.

The function $f(x) = 3x$ describes the relationship in the table.

x		$f(x)$
3	$9 = 3 \cdot 3$	9
4	$12 = 3 \cdot 4$	12
5	$15 = 3 \cdot 5$	15

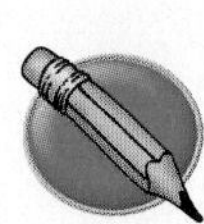

Describe the relationship between the input and the output values for the function *f*(*x*) = 2*x* – 5.

____

____

Practice

Write a function for each table of values.

1.

Temperature	Pressure (pounds per square inch, psi)
10°F	30 psi
20°F	40 psi
30°F	50 psi
$f(x) = x + 20$	

2.

Number rolled	Score
5	20 points
6	24 points
7	28 points
$f(x) =$	

3.

Travel time	Ticket price
1 hour	$125
2 hours	$175
3 hours	$225
$f(x) =$	

4.

Correct answers	Points
10	10
9	9
8	8
$f(x) =$	

Evaluate each function to solve the problem.

5. Use the function for temperature and pressure to find $f(50)$.

   $f(50) = 50 + 20 = 70$

6. Use the function for number rolled and score to find $f(2)$.

7. Use the function for travel time and ticket price to find $f(6)$.

8. Use the function for correct answers and points to find $f(7)$.

Think about a function that relates time and distance. For example, $m = 60t$ gives the number of miles, m, driven in an hour when t is given in hours. Explain why it is practical to limit the domain and range of this function to positive numbers.

Lesson 12

Functions and Graphs

The input and output values for a function can be used as ordered pairs to graph the function.

Make a table of values of the function.

Model 1 **Complete the table for $f(x) = \frac{1}{2}x$.**

x	$f(x)$	$(x, f(x))$
⁻4	⁻2	(⁻4,⁻2)
⁻2	⁻1	(⁻2,⁻1)
0		
2		
4		

Note that $f(x)$ is another name for y in an ordered pair. When you graph a function, the y-axis may be labeled with $f(x)$.

Model 2 **Graph the function $f(x) = \frac{1}{2}x$ using the table of values.**

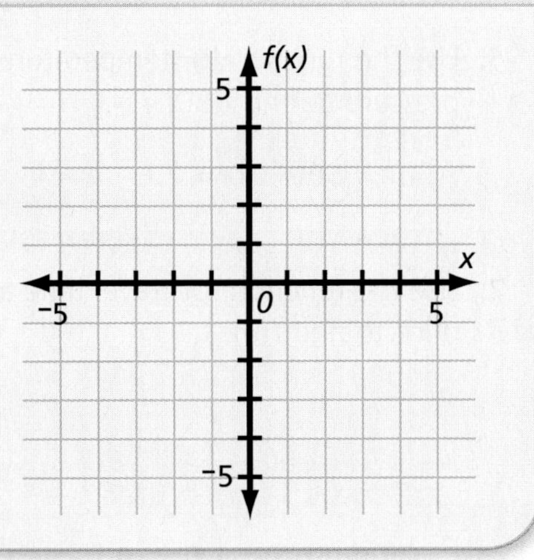

Explain how to select values for x to make a graph for $f(x) = {}^{-}2x + 3$.

____

____

Practice

Complete the table for each function.

1.

$f(x) = x + 2$		
x	$f(x)$	$(x, f(x))$
$^-3$	$^-1$	$(^-3, ^-1)$
$^-1$		
1		

2.

$f(x) = {}^-2x$		
x	$f(x)$	$(x, f(x))$
2		
1		
0		

3.

$f(x) = \frac{2}{3}x + 2$		
x	$f(x)$	$(x, f(x))$
$^-3$		
3		
0		

4.

$f(x) = 3x + 1$		
x	$f(x)$	$(x, f(x))$
$^-1$		
0		
1		

Graph each of the functions using the tables from problems 3 and 4.

5. $f(x) = \frac{2}{3}x + 2$

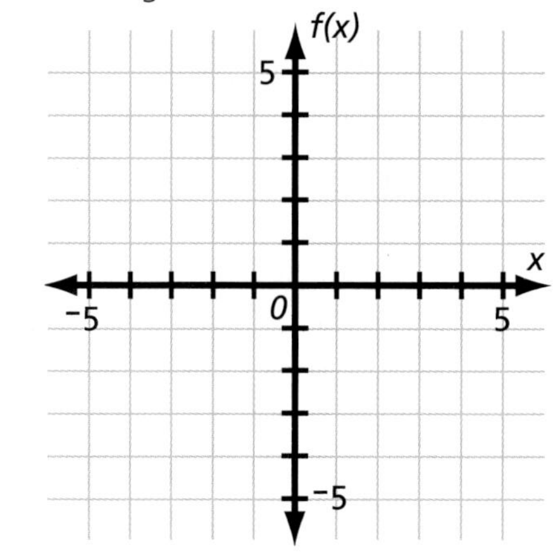

6. $f(x) = 3x + 1$

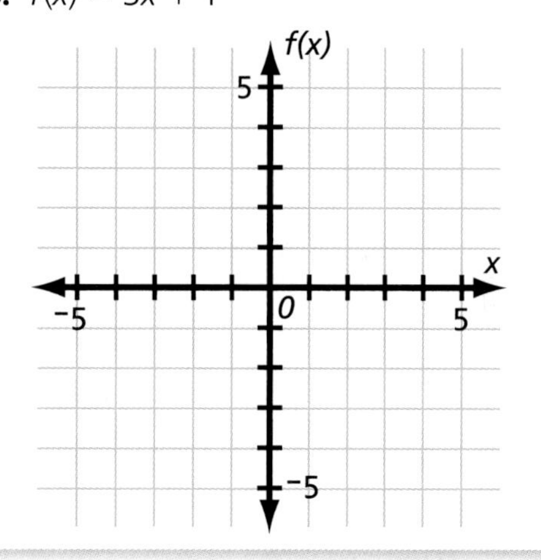

Explain why $^-3$, 0, and 3 would be good choices to put in a table of values for the function $f(x) = \frac{1}{3}x + 2$.

____

____

____

____

Strength Builder

Fun with Linear Equations

You can play *Over and Under* individually or in teams.

Materials

You will need graph paper, a two-colored counter, and two number cubes.

Getting Ready to Play

1. Each player or team of players creates one linear equation. Use any positive or negative number between $^{-}12$ and 12 for the slope and *y*-intercept. Then graph the equation.
2. Choose a student to be the announcer.
3. Decide which color represents positive and which represents negative.
4. Each player or team of players makes a scorecard like the sample below.

Name:		**Equation:**	
Round	**(*x*, *y*)**	**Over graph**	**On graph**
1.			
2.			
3.			
4.			
5.			
6.			
7.			
8.			
9.			
10.			

How to Play

1. The announcer flips the counter and rolls the number cubes to find the coordinates for ten ordered pairs. The counter determines whether a coordinate is positive or negative according to the color.
2. If the coordinates are above the graph of your equation, write $^{+}1$ in the column labeled Over Graph. If the coordinates are a solution to your equation, write $^{+}2$ in the column labeled On Graph. If the coordinates are below the graph of your equation, you do not earn any points.
3. Repeat for the remaining ordered pairs.
4. At the end of round 10, all players or teams check their work.
5. The player or team with the most points wins.

Variations

1. Graph inequalities. If the announcer calls out coordinates in the solution set of your inequality, you earn a point.
2. At the end of round 10, players or teams swap scorecards and check each other's work.
3. Use a graphing calculator to graph equations. Verify points on the line using the table of solutions on the calculator.

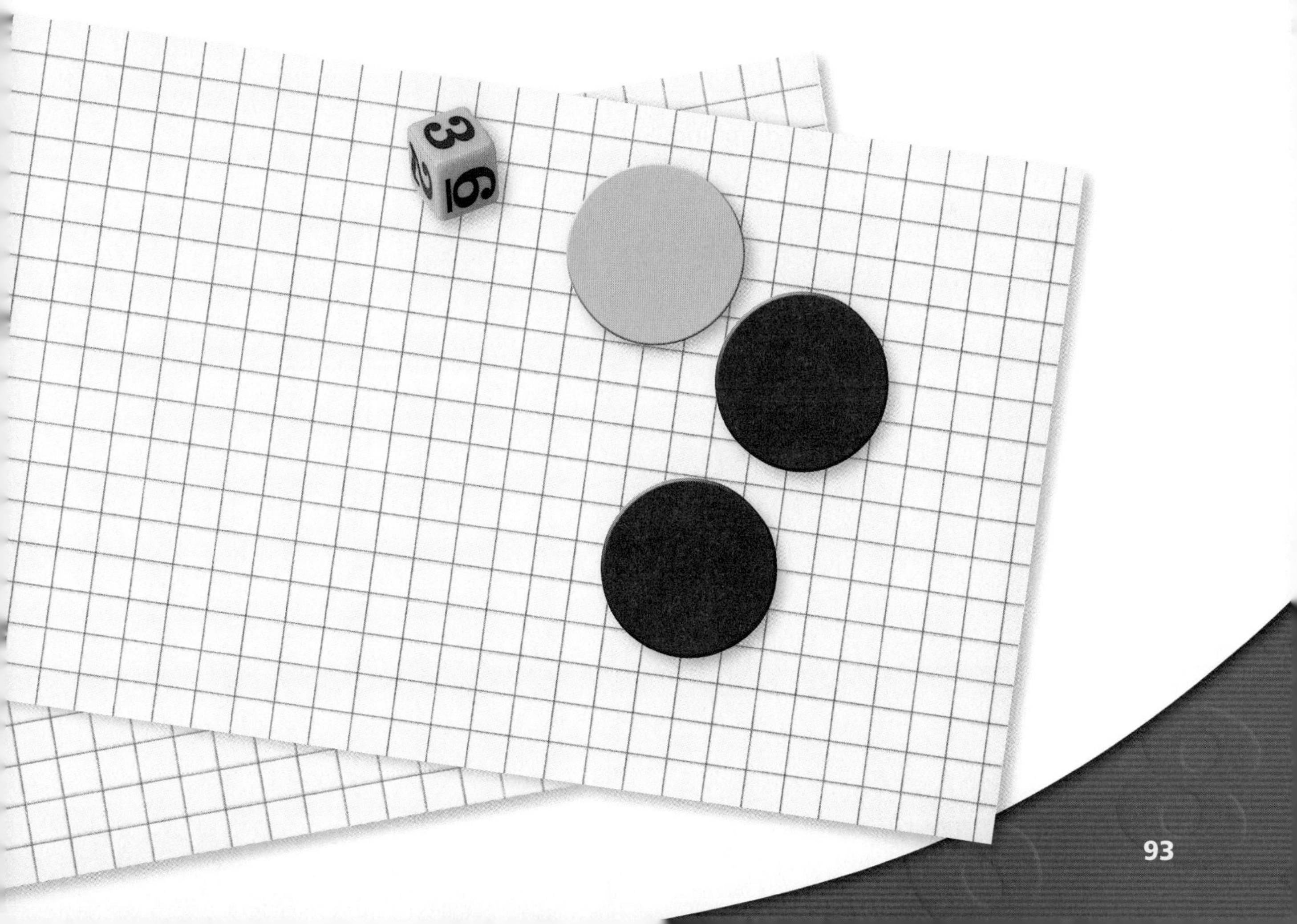

Unit 3 Review

Complete each table.

1.

x	$y = 3x - 2$	(x, y)
$^-1$		
0		
1		

2.

x	$y = x + 2$	(x, y)
2		
4		
7		

Solve.

3. Use the two points to determine the slope of line AB.
   $A(2, {^-5})$ $B(3, {^-1})$
   $m =$

4. Identify the slope and y-intercept of the line given by $y = \frac{1}{4}x + 4$.
   $m =$; y-intercept $=$

5. Write the equation of a line with slope $= {^-5}$ and y-intercept $= {^-1}$.

6. Find the constant of variation, k, where $y = 30$, $x = 15$ and y varies directly as x.
   $k =$

7. Determine whether $(1, {^-3})$ is a solution of $y = {^-x} - 1$.

8. What is the y-intercept of the graph of $y = 4x - 2$?
   y-intercept $=$

Solve the system of equations by graphing. State the solution.

9. $y = {^-x} + 1$
   $y = x + 3$
   solution:

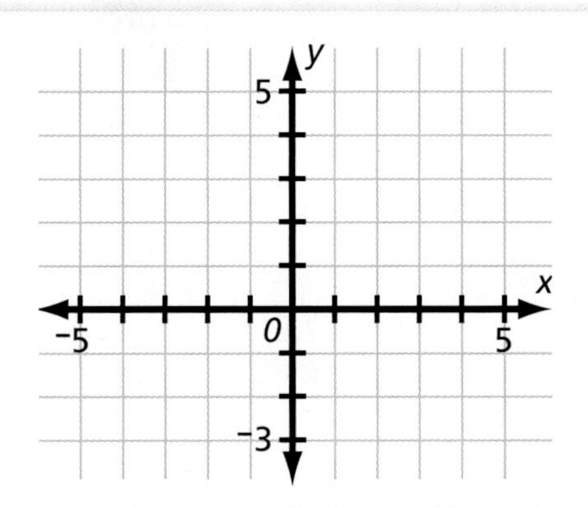

10. State the domain and range of the function $f(x) = \frac{1}{2}x + 5$.

11. Evaluate the function $f(x) = {^-3}x + 1$ for $x = 4$.

12. Evaluate the function $f(x) = {^-\frac{1}{2}}x$ for $x = {^-8}$.

Unit 4

Polynomials

Did you know

that most sunflowers have 34 petals?

There are many connections between patterns in mathematics and patterns in nature. A buttercup has 5 petals, and a daisy has either 55 or 89 petals. These numbers are modeled by the Fibonacci sequence. Every number in the sequence is the sum of the two numbers before it. What are the next five numbers in this sequence?

1, 1, 2, 3, 5, 8, 13, 21, 34

Lesson 1 Polynomials

Constants, variables, and products of constants and variables are **monomials**. Expressions with radical signs, or division by variables are not monomials. A **polynomial** is the sum or difference of monomials.

A **binomial** is the sum or difference of two monomials. A **trinomial** is the sum or difference of three monomials.

Constant	Number	6 9 2 14
Monomial	One term	$6a$ $9yz$ $2x$ $14s^2$
Binomial	Two terms	$6a + 9yz$ $2x - 14s^2$
Trinomial	Three terms	$6a + 9yz - 2x$ $14s^2 + 6a + 2$
Polynomial	Two or more terms	$6a + 9yz$ $6a + 9yz + 2 + 14s^2$

In **standard form**, all like terms have been combined. The order of terms begins with the term with the greatest exponent.

Model 1 **Write $x - 3x^2 + 9 - 2x + 5x^2$ in standard form.**

$^-3x^2 + 5x^2 + x + 9 - 2x =$ **Think:** The greatest exponent is 2, so the terms with squares are first.

$^-3x^2 + 5x^2 + x - 2x + 9 =$ Use the Commutative Property to change the order again. The x terms are next and the constant is last.

______ $- x +$ ______ $=$ Combine like terms.

$2x^2 - x + 9 =$ **Check:** exponent of 2, exponent of 1, constant

Explain the difference between a monomial and a binomial.

____

____

____

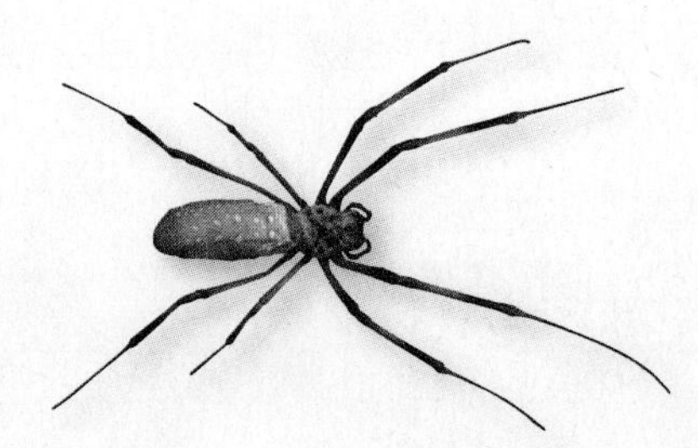

Practice

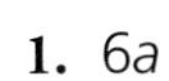

Identify each expression as *monomial, binomial,* or *trinomial.*

1. $6a$

   monomial

2. $\frac{5t}{2} - 5$
3. $s - 4t + 8$
4. $^{-}4x + 12$
5. xyz
6. ^{-}b
7. $^{-}10n^2 + 6$
8. $s^4 + s - 73$
9. $y^3 + y^2 + 3y$

If the polynomial is in standard form write *standard form*. Otherwise rewrite the expression in standard form.

10. $3 + 8n$

    $8n + 3$

11. $x^2 + 3x + 2$
12. $3y^4 - 10 + y^3 + 2y$
13. $u^5 + u^4$
14. $^{-}d + 4 + d^2$
15. $2x + 5$
16. $9p^3 - 2p$
17. $6k^3 - 9 + 3k^2$
18. $9 + 8q^3 + 7q^2 + 6q$

Combine like terms and write the polynomial in standard form.

19. $4p^2 - 8 + 5p^2 + 2p$

    $9p^2 + 2p - 8$

20. $12b - 9b + 1 + 3b^2$
21. $^{-}4 + 5a^3 + 7a^2 + 6a^3$
22. $3w + w^2 - 8w - 4w^2$
23. $2 + q^4 + 2q^2 - 7$
24. $6x^4 - x + 3x^4 - 5x$

Explain which terms in the polynomial $7x^3 - xy + 3x^2y - 5x^2 + 4xy$ can be combined.

Lesson 2

Addition of Polynomials

To add two polynomials means to add the like terms of the polynomials. Use the properties of addition to group and add like terms. Recall that like terms have the same variables raised to the same power.

Like terms	$4y$ and ^-7y	$7cd$ and cd	$9w^3$ and $^-5w^3$
Unlike terms	$4y$ and $4z$	$7c$ and $7cd$	$9w^3$ and $^-5w^2$

Model 1

Add $x + 3$ and $2x + 11$.

$(x + 3) + (2x + 11) =$ — **Think:** Remove parentheses and rearrange like terms together.

$x + 2x + 3 + 11 =$ — Use the Commutative Property to rearrange terms.

$3x + 14$ — Combine like terms.

Add $4x^2 + x - 7$ and $2x + 9$.

$(4x^2 + x - 7) + (2x + 9) =$

$4x^2 + x + 2x + (^-7) + 9 =$ — Use the Commutative Property to rearrange terms.

$4x^2 +$ ______ $+$ ______ — Combine like terms.

Model 2

Add $2a^2 + 3ab + 8$ and $5a + 4ab + 1$.

$(2a^2 + 3ab + 8) + (5a + 4ab + 1) =$

$2a^2 + 3ab + 4ab + 5a + 8 + 1 =$ — Use the Commutative Property to rearrange terms.

$2a^2 +$ ______ $+ 5a +$ ______ — Combine like terms.

How many terms will the sum of $(3x + xy + 1)$ and $(x + y + 3)$ have? Explain your answer.

____

____

Practice

Add the polynomials.

1. $a + 3$ and $2a + 9$

   $(a + 3) + (2a + 9) =$
   $a + 2a + 3 + 9 =$
   $3a + 12$

2. $n + 5m + 1$ and $3n + m + 7$

3. $^{-}4c + c^2 + 8$ and $3c + 1$

4. $2d + 4cd + 2$ and $d + cd + 1$

5. $b + 10$ and $10 + b + b^2$

6. $h^2 + h + 3$ and $^{-}3h + 2$

7. $5t^2 + 2$ and $2t + 5 - 3t^2$

8. $v - 8 + w$ and $3w + 7v$

9. $^{-}7u + 3$ and $5uy + 4 + 5u$

10. $4z - 2$ and $z^2 - 8z$

11. $2w^3 - 5w^2 + 6$ and $w^2 + 4w - 8$

12. $^{-}10 + 3ab$ and $a^2 - ab$

Explain why the sum of $x^2 + 5x - 3$ and $x^2 - 5x + 3$ is a monomial.

Lesson 3

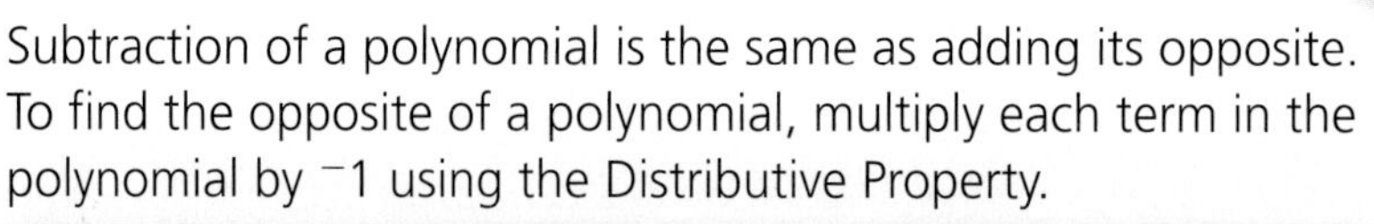

Subtraction of Polynomials

Subtraction of a polynomial is the same as adding its opposite. To find the opposite of a polynomial, multiply each term in the polynomial by $^-1$ using the Distributive Property.

Polynomial	Opposite Polynomial
$a + b$	$^-(a + b) = (^-1)(a + b) = {}^-a - b$
$x - y$	$^-(x - y) = (^-1)(x - y) = {}^-x + y$
$2c^2 - 4$	$^-(2c^2 - 4) = (^-1)(2c^2 - 4) = {}^-2c^2 + 4$

Write the opposite of each polynomial.

Model 1

$x + 6$

Distribute $^-1$ to each term.

$^-(x + 6) =$

$(^-1)(x + 6) =$

$^-x - 6$

$3x^2 - 3x + 5$

Distribute $^-1$ to each term.

$^-(3x^2 - 3x + 5) =$

$(^-1)(3x^2 - 3x + 5) =$

$^-3x^2 +$ _______ $-$ _______

Subtract by adding the opposite of the polynomial being subtracted.

Model 2

$7m^2 + 3m - (2m^2 + m) =$	**Think:** Subtraction is the same as adding the opposite.
$7m^2 + 3m + (^-1)(2m^2 + m) =$	Change the $-$ to $+$ and multiply the polynomial being subtracted by $^-1$.
$7m^2 + 3m + {}^-2m^2 - m =$	Use the Distributive Property.
$7m^2 - 2m^2 + 3m + {}^-m =$	Use the Commutative Property.
$5m^2 +$ _______	Combine terms.

Why is it correct to say that only like terms of polynomials can be subtracted?

____

____

____

Practice

Write the opposite of each polynomial.

1. $2b - 8$

$^{-}1(2b - 8) =$
$^{-}2b + 8$

2. $3p + 4$

3. $3q^2 - 5q + 4$

4. $6c^2 - 6c + 2$

5. $5d - 5$

6. $10r^2 + 7r - 3$

Subtract each of the following polynomials.

7. $4x + 2xy - (xy + 5) =$

$4x + 2xy - xy - 5 =$
$4x + xy - 5$

8. $^{-}8w^2 - (2w + 7) =$

9. $12 - (^{-}6z + 8) =$

10. $3 - (n + 1) =$

11. $5w^2 - (2w - 1) =$

12. $(a + 3) - (4a + 5) =$

13. $(4k^2 + 3k) - (3k^2 + 3k) =$

14. $(4s - 5) - (s + 9) =$

15. $(cd - 2c) - (5cd + 2) =$

16. $10v^2 - (8v^2 + 4v) =$

17. $^{-}7b - (^{-}2b + 1) =$

18. $7r - (r + rs + 2) =$

Write a subtraction of polynomials problem with a difference of zero.

____

____

Multiplication of Polynomials by Monomials

To multiply a polynomial by a monomial, multiply the coefficients and use the rules of exponents.

Examples:

$4n^2 \cdot 5n^3 = 20n^5$	Add exponents in multiplication when the bases are the same.
$^-n \cdot 6n^3 = {^-6}n^4$	Remember that n means n^1.

To multiply a binomial by a monomial, use the Distributive Property and follow the rules of exponents.

Model 1

$3x(2x + 8) =$

$3x \cdot$ ______ $+ 3x \cdot$ ______ $=$

______ + ______

If the polynomial is a trinomial, multiply all three terms of the trinomial by the monomial.

Model 2

$5x(2x^2 - 8x + 2) =$

$(5x \cdot 2x^2) + (5x \cdot {^-8}x) + (5x \cdot$ ______$) =$

$10x^3 +$ ______ + ______

When multiplying, the product of terms with matching signs is positive. The product of terms with unlike signs is negative.

Model 3

$^-3x^4(x^2 + x + 5) =$

$({^-3}x^4 \cdot$ ______$) + ({^-3}x^4 \cdot$ ______$) + ({^-3}x^4 \cdot$ ______$) =$

______ + ______ + ______

Explain why the product of a monomial and a polynomial may have a higher exponent than either the polynomial or the monomial.

____

____

____

Practice

Multiply.

1. $4(a + 8) =$ $4a + 32$	**2.** $6(p - 10) =$	**3.** $3(4r^2 + 2r) =$
4. $^-5b(3b^2 + 6) =$	**5.** $2s(s^2 + 5s) =$	**6.** $7w^2(w^2 + 2w) =$
7. $^-2(8n^3 - 5n) =$	**8.** $9(5q^2 + 3q) =$	**9.** $10y(3y^3 - 4) =$

Multiply.

10. $2z(z^2 + z - 8) =$ $2z^3 + 2z^2 - 16z$	**11.** $^-k(3k^2 + 6k - 1) =$	**12.** $5(n^2 + n + 2) =$
13. $w^4(2w^2 + 5w - 12) =$	**14.** $20b^3(b^2 - 3b + 9) =$	**15.** $^-a(9a^2 - 2a + 5) =$
16. $5n(3n^2 - n) =$	**17.** $12a^2(3a + 1) =$	**18.** $^-1(^-4b + 7) =$

Explain how you would simplify the problem $2x(x^2 + 3x - 1) - 2x(x + 7)$ and give the result.

____

____

____

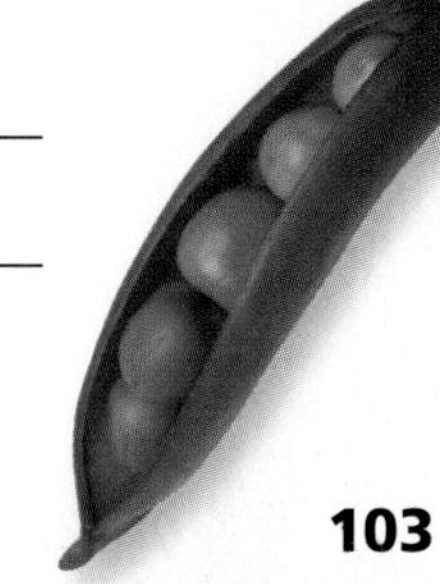

Division of Polynomials by Monomials

To divide a polynomial by a monomial, divide each term of the polynomial by the monomial. Follow the rules for division of expressions with exponents.

Examples:

$\frac{10a^5}{5a^2} = 2a^3$	Subtract the exponent in the denominator from the exponent in the numerator when the bases are the same.
$\frac{a^4}{-4a} = -\frac{1}{4}a^3$	Remember that a means a^1 when subtracting exponents.

Use the rules for division with exponents to divide a polynomial by a monomial.

Model 1

$\frac{22x^3 + 22x^2 - 11x}{11} =$ **Think:** Write the expression as separate fractions.

$\frac{22x^3}{11} + \frac{22x^2}{11} - \frac{\_\_\_}{11} =$ Divide each term in the numerator by the denominator.

______ + ______ − ______ Simplify each fraction.

Another way to divide a polynomial by a monomial is to use factoring first. **Factor the greatest common factor from all the terms of the numerator. Then simplify.**

Model 2

$\frac{8x^4 + 4x^2}{2x^2} =$ **Think:** The greatest common factor of $8x^4$ and $4x^2$ is $4x^2$.

$\frac{4x^2(2x^2 + 1)}{2x^2} =$ Use $4x^2$ as a factor.

______$(2x^2 + 1) =$ Simplify $4x^2$ and the denominator.

______ + ______ Use the Distributive Property.

Explain how to divide the problem in Model 1 by finding the greatest common factor of the numerator first. Compare your answer to Model 1.

____

____

____

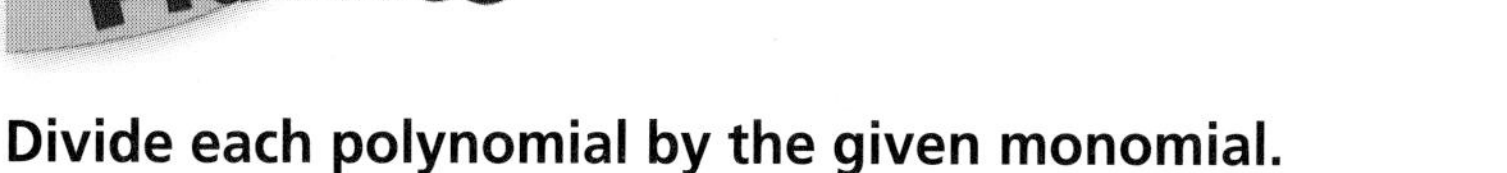

Practice

Divide each polynomial by the given monomial.

1. $\frac{3b + 12}{3} =$

   $b + 4$

2. $\frac{8z - 32}{4} =$

3. $\frac{4n^3 + n^2}{n^2} =$

4. $\frac{7w^2 + 7w - 14}{^-7} =$

5. $\frac{9y^2 - 3}{3} =$

6. $\frac{8z^4 - 4z^2 + 8}{4} =$

Divide each polynomial by factoring the greatest factor from the numerator first.

7. $\frac{6n^2 - 3n + 6}{3} =$

   $\frac{3(2n^2 - n + 2)}{3} =$

   $2n^2 - n + 2$

8. $\frac{5a^3 + 5a^2 + 10a}{5a} =$

9. $\frac{40t^2 - 10t}{5t} =$

10. $\frac{16b^2 - 4}{^-4} =$

11. $\frac{4n^3 + 2}{2} =$

12. $\frac{100x^2 + 10x - 5}{5}$

13. $\frac{30t - 10}{5} =$

14. $\frac{3rs + s}{s} =$

15. $\frac{9n^3 + 6n + 9}{3} =$

Explain the error a student made by writing $\frac{4z^3 + 8z}{2z} = 2z^2 + 4z$.

____

____

Monomials and Powers

To raise a monomial to a power, raise each factor in the monomial to the power indicated. All the factors within a set of parentheses should be raised to the power indicated.

Examples:

$(3x)^2$	$(5cd)^3$	$^-8(w^2)^3$
Raise $3x$ to the 2nd power.	Raise $5cd$ to the 3rd power.	Raise w^2 to the 3rd power; multiply by $^-8$ once.

Simplify.

Model 1

$(3x)^2 =$	Raise $3x$ to the 2nd power.
$3x \cdot 3x =$	$3x$ is a factor two times.
$3 \cdot 3 \cdot x \cdot$ ______ $=$	Use Commutative Property to rearrange terms.
______	Multiply coefficients; multiply variables.
$(5cd)^3 =$	Raise $5cd$ to the 3rd power.
$5cd \cdot 5cd \cdot 5cd =$	$5cd$ is a factor three times.
$5^3 \cdot$ ______ $\cdot$ ______ $=$	Multiply coefficients; multiply variables.
______	

Simplify. Use the factor outside the parentheses only once.

Model 2

$^-8(w^2)^3 =$	Raise w^2 to the 3rd power.
$^-8 \cdot w^2 \cdot w^2 \cdot$ ______ $=$	$^-8$ is a factor once, w^2 is a factor three times.
$^-8 \cdot$ ______ $=$	Multiply variables by adding the exponents.
______	

Explain why $^-(x^2)^2$ and $(^-x^2)^2$ do not have the same result when they are simplified.

____

____

Practice

Simplify.

1. $(4x)^3 =$

   $64x^3$

2. $(2w)^4 =$

3. $(ab)^3 =$

4. $(2z)^5 =$

5. $(^-7c)^2 =$

6. $(5y)^4 =$

7. $(2x^3)^2 =$

8. $(3y^7)^2 =$

9. $(^-kn^2)^2 =$

10. $(6w^2z^3)^2 =$

11. $(c^2df)^4 =$

12. $(3y^5)^4 =$

Simplify.

13. $3(x^2)^3 =$

    $3x^6$

14. $^-4(q^2)^3 =$

15. $10(p^4b)^2 =$

16. $2(p^4)^2 =$

17. $^-3(2b)^3 =$

18. $3(9n)^2 =$

Explain the steps needed to simplify the expression $^-8(^-2x^3)^2$ and give the result.

___

___

___

Binomials and Multiplication

To multiply two binomials, use the Distributive Property. The property must be applied twice.

Multiply $(x + 2)(x + 7)$ using the Distributive Property.

Model 1

First multiply $(x + 7)$ by x. Then multiply $(x + 7)$ by 2. Finally, combine any like terms.

$$(x + 2)(x + 7) =$$

$(x + 2)(x + 7) =$	Multiply by x, then by 2.
$x(x + 7) + 2(x + 7) =$	Add the results, since 2 is positive.
$(x \cdot x) + (x \cdot 7) + (2 \cdot x) + (2 \cdot 7) =$	Distribute x. Then distribute 2.
$x^2 + 7x +$ ______ $+ 14 =$	Multiply.
$x^2 +$ ______ $+ 14$	Combine like terms.

Multiply $(3x + 5)(x - 2)$.

Model 2

First multiply $(x - 2)$ by $3x$. Then multiply $(x - 2)$ by 5.

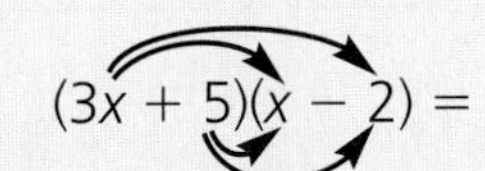

$$(3x + 5)(x - 2) =$$

$(3x + 5)(x - 2) =$	Multiply by $3x$, then by 5.
$3x(x - 2) + 5(x - 2) =$	Add the results, since 5 is positive.
$(3x \cdot x) + (3x \cdot {}^{-}2) + (5 \cdot x) + (5 \cdot {}^{-}2) =$	Distribute $3x$. Then distribute 5.
$3x^2 +$ ______ $+$ ______ $- 10 =$	Multiply.
____________________	Combine like terms.

Explain how to combine like terms when you multiply $(a + b)(a + b)$.

____

____

Practice

Use the Distributive Property to multiply each pair of binomials.

1. $(p + 3)(p + 2) =$

$p(p + 2) + 3(p + 2) =$
$p^2 + 2p + 3p + 6 =$
$p^2 + 5p + 6$

2. $(k + 7)(k + 9) =$

3. $(r + 5)(r + 3) =$

4. $(w + 5)(w - 8) =$

5. $(z + 5)(z - 4) =$

6. $(x + 8)(x + 1) =$

Multiply each pair of binomials.

7. $(a + 10)(a + 2) =$

$a^2 + 2a + 10a + 20 =$
$a^2 + 12a + 20$

8. $(2b + 4)(b + 7) =$

9. $(w + 3)(w + 4) =$

10. $(6n + 1)(n - 2) =$

11. $(c + 4)(5c - 4) =$

12. $(b + 4)(b + 7) =$

13. $(k + 5)(10k + 6) =$

14. $(7n + 3)(n - 1) =$

15. $(a + 9)(a - 6) =$

Multiply $(x + 3)(x - 3)$. Explain why the product is a binomial.

____

____

Lesson 8

Square of the Sum of Two Terms

A sum of two terms is a binomial. To square a binomial means to multiply the binomial by itself.

Square each binomial.

Model 1

$(x + 3)^2 =$

$(x + 3)(x + 3) =$

$x(x + 3) + 3(x + 3) =$

$x^2 + 3x + 3x + 9 =$

$x^2 +$ ______ $+ 9$

$(2x + 5)^2 =$

$(2x + 5)(2x + 5) =$

$2x(2x + 5) + 5(2x + 5) =$

$4x^2 + 10x +$ ______ $+$ ______ $=$

$4x^2 +$ ______ $+$ ______

When a binomial is squared, there is a pattern to the terms in the product.

Multiply to see the pattern for squaring the sum of two terms.

Model 2

$(x + a)^2 =$ (__________)(__________) $=$

$x(x + a) + a(x + a) =$

first term squared	plus	twice the product of the two terms	plus	last term squared
$x \cdot x$	+	$x \cdot a + x \cdot a$	+	$a \cdot a$
x^2	+	$2ax$	+	a^2

Use the pattern to square each binomial.

Model 3

$(x + 6)^2 =$

$x^2 + 2 \cdot 6 \cdot x +$ ______ $=$

____

$(3x + 7)^2 =$

$9x^2 + 2 \cdot 7 \cdot 3x +$ ______ $=$

____

In the pattern $(x + a)(x + a) = x^2 + 2ax + a^2$, why must a^2 always be positive?

____

____

Practice

Square each binomial.

1. $(b + 4)^2 =$ $b^2 + 4b + 4b + 16 =$ $b^2 + 8b + 16$	2. $(z + 10)^2 =$	3. $(w + 8)^2 =$
4. $(3c + 5)^2 =$	5. $(5a + 4)^2 =$	6. $(k + 6)^2 =$
7. $(w + 1)^2 =$	8. $(2d + 3)^2 =$	9. $(2n + 6)^2 =$

Multiply. Follow the pattern for squaring the sum of two terms.

10. $(h + 13)^2 =$ $h^2 + 26h + 169$	11. $(m + 5)^2 =$	12. $(4y + 3)^2 =$
13. $(2x + 1)^2 =$	14. $(d + 11)^2 =$	15. $(z + 7)^2 =$

Explain how to follow the pattern $(x + a)(x + a) = x^2 + 2ax + a^2$ to square $(5y + 2)$.

____

____

____

Square of the Difference of Two Terms

The difference of two terms is a binomial. To square a binomial means to multiply the binomial by itself.

Square each binomial.

Model 1

$(x - 7)^2 =$

$(x - 7)(x - 7) =$

$x(x - 7) - 7(x - 7) =$

$x^2 - 7x - 7x + 49 =$

$x^2 -$ ______ $+ 49$

$(4x - 1)^2 =$

$(4x - 1)(4x - 1) =$

$4x(4x - 1) - 1(4x - 1) =$

$16x^2 - 4x -$ ______ $+$ ______ $=$

$16x^2 -$ ______ $+$ ______

When the difference of two terms is squared, there is a pattern to the terms in the product.

Multiply to see the pattern for squaring the difference of two terms.

Model 2

$(x - a)^2 = ($__________$)($__________$) =$

$x(x - a) - a(x - a) =$

first term squared	minus	twice the product of the two terms	plus	last term squared
$x \cdot x$	$-$	$x \cdot a - x \cdot a$	$+$	$a \cdot a$
x^2	$-$	$2ax$	$+$	a^2

Use the pattern to square each binomial.

Model 3

$(x - 9)^2 =$

$x^2 - 2 \cdot$ ______ $\cdot$ ______ $+ 81 =$

____

$(2x - 5)^2 =$

$4x^2 - 2 \cdot$ ______ $\cdot$ ______ $+ 25 =$

____

Explain why the second term in the product $(x - 15)(x - 15)$ must have a coefficient that is even.

____

____

Practice

Square each binomial.

1. $(b - 4)^2 =$

   $b^2 - 4b - 4b + 16 =$
   $b^2 - 8b + 16$

2. $(z - 12)^2 =$

3. $(h - 1)^2 =$

4. $(6c - 2)^2 =$

5. $(q - 8)^2 =$

6. $(k - 20)^2 =$

7. $(w - 11)^2 =$

8. $(2h - 5)^2 =$

9. $(y - 16)^2 =$

Multiply. Follow the pattern for squaring a difference of two terms.

10. $(x - 10)^2 =$

    $x^2 - 20x + 100$

11. $(5m - 1)^2 =$

12. $(y - 7)^2 =$

13. $(3x - 2)^2 =$

14. $(c - 2)^2 =$

15. $(n - 6)^2 =$

Explain how the two patterns for squaring binomials are alike and how they are different.

Lesson 10

Perfect Square Trinomials

A trinomial that results from squaring a binomial is a **perfect square trinomial**. A perfect square trinomial follows a pattern. The middle term is two times the square root of the first term, times the square root of the last term.

Square of the sum of two terms	Square of the difference of two terms
$(x + a)^2 = x^2 + 2ax + a^2$	$(x - a)^2 = x^2 - 2ax + a^2$
Example: $(x + 9)^2 = x^2 + 18x + 81$	**Example:** $(x - 6)^2 = x^2 - 12x + 36$

Use the patterns to factor each perfect square trinomial.

Model 1

$x^2 + 10x + 25$

square root of x^2 = ______ square root of 25 = ______

$2 \cdot x \cdot 5$ = ______

Use the square roots of the first and last terms to write the binomial of this perfect square trinomial.

$x^2 \cdot 10x + 25 = (\_\_\_\_\_\_ + \_\_\_\_\_\_)^2$

Model 2

$x^2 + 20x + 100 =$

$x^2 + (2 \cdot 10x) +$ ______ $=$

$(x + 10)(x + 10) =$

$(x +$ ______$)^2$

$x^2 - 14x + 49 =$

$x^2 - (2 \cdot$ ______$) + 7^2 =$

$(x - 7)(x - 7) =$

$(x -$ ______$)^2$

Is $x^2 + 10x + 16$ a perfect square trinomial? Why or why not?

____

____

____

____

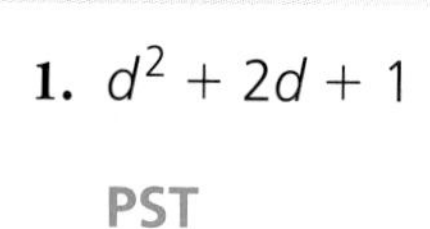

Practice

If the polynomial is a perfect square trinomial write *PST.* Otherwise explain why it is not a perfect square trinomial.

1. $d^2 + 2d + 1$

PST

2. $m^2 + 4m + 9$

3. $r^2 - 10r + 25$

4. $p^2 - 20p + 100$

5. $v^2 + 8v + 4$

6. $b^2 - 12b + 16$

7. $s^2 + 24s + 144$

8. $q^2 + 28q + 196$

Factor each perfect square trinomial.

9. $x^2 - 4x + 4$

$(x - 2)(x - 2) = (x - 2)^2$

10. $m^2 + 22m + 121$

11. $n^2 + 40n + 400$

12. $p^2 - 30p + 225$

13. $w^2 + 18w + 81$

14. $q^2 - 12p + 36$

Explain how to find a value for k so that $x^2 + 16x + k$ is a perfect square.

Difference of Squares

Some binomials follow a pattern called **difference of squares**. Consider the following multiplication problems.

$(y + 5)(y - 5) =$	$(x + 1)(x - 1) =$	$(2w + 9)(2w - 9) =$
$y^2 - 5y + 5y - 25 =$	$x^2 - x + x - 1 =$	$4w^2 - 18w + 18w - 81 =$
$y^2 - 25$	$x^2 - 1$	$4w^2 - 81$

Each of these products is a difference of squares. The first and last terms are perfect squares with a subtraction sign between them. It is the subtraction sign that gives the pattern the name *difference*.

A binomial that is a difference of squares factors in a special way. **Factor each difference of squares.**

Model 1

$x^2 - 49 =$

$(x + 7)(x - 7)$

Use the square roots to place terms in the binomials. Use one addition and one subtraction sign.

$x^2 - 64 =$

$(x +$ ______ $)(x -$ ______ $)$

Use the square roots to place terms in the binomials. Use one addition and one subtraction sign.

Model 2

$4x^2 - 25 =$

$(2x + 5)(2x - 5)$

Use the square roots to place terms in the binomials. Use one addition and one subtraction sign.

$(2x + 5)(2x - 5) =$

$2x \cdot 2x - 10x + 10x - 25 =$

$4x^2$ ______

Check the factoring by multiplying.

Explain how to decide if a polynomial is a difference of squares.

____

____

____

____

____

Practice

Factor each difference of squares.

1. $x^2 - 100 =$

$(x + 10)(x - 10)$

2. $y^2 - 4 =$

3. $k^2 - 121 =$

4. $z^2 - 36 =$

5. $w^2 - 400 =$

6. $h^2 - 1 =$

7. $c^2 - 9 =$

8. $w^2 - 144 =$

9. $y^2 - 169 =$

Factor each difference of squares.

10. $9x^2 - 25 =$

$(3x + 5)(3x - 5)$

11. $9c^2 - 100 =$

12. $36z^2 - 1 =$

13. $49x^2 - 4 =$

14. $100n^2 - 49 =$

15. $25h^2 - 144 =$

16. $400m^2 - 49 =$

17. $100w^2 - 1 =$

18. $225b^2 - 4 =$

Explain why it is correct to factor $4x^2 - 25$ as either $(2x + 5)(2x - 5)$ or $(2x - 5)(2x + 5)$.

Lesson 12 Solutions to Equations

Use the Zero Property of Multiplication to solve for the variable x in the equation $2x = 0$.

Zero Property of Multiplication
$a \cdot 0 = 0 \cdot a = 0$

The property states that when the product is zero, one of the factors must equal zero.

In $2x = 0$, x must equal zero.

When an equation is written in factored form and set equal to zero, the Zero Property of Multiplication can be used to solve it.

Factor the equation and then solve for x.

Model 1

$x^2 + 12x + 36 = 0$	The polynomial is a perfect square trinomial.
$(x + 6)(x + 6) = 0$	The equation is now in factored form.
$x + 6 = 0$	When the product is zero, at least one of the factors must be zero. Both factors are the same.
$x = {}^{-}6$	The equation has one solution, $x = {}^{-}6$.

Factor the equation and then solve for x.

Model 2

$x^2 - 64 = 0$	The polynomial is a difference of squares.
$(x + 8)(x - \_\_\_\_\_\_) = 0$	The equation is now in factored form.
$x + 8 = 0$ or $x - 8 = 0$	When the product is zero, at least one of the factors must be zero.
$x = {}^{-}8$ or $x = \_\_\_\_\_\_$	The equation has two solutions, $x = {}^{-}8$ and $x = 8$.

You cannot use the Zero Property of Multiplication to solve an equation unless the equation is set equal to zero. Explain the steps necessary to solve $x^2 + 12x + 40 = {}^{-}2x - 9$ and give the result.

____

____

____

Practice

Solve each equation.

1. $x^2 + 18x + 81 = 0$

   $(x + 9)(x + 9) = 0$
   $(x + 9) = 0$
   $x = ^-9$

2. $n^2 + 2n + 1 = 0$

3. $w^2 - 10w + 25 = 0$

4. $y^2 + 24y + 144 = 0$

5. $x^2 - 4x + 4 = 0$

6. $b^2 + 16b + 64 = 0$

Solve each equation.

7. $x^2 - 100 = 0$

   $(x - 10)(x + 10) = 0$
   $(x - 10) = 0; (x + 10) = 0$
   $x = 10; x = ^-10$

8. $y^2 - 25 = 0$

9. $w^2 - 144 = 0$

10. $c^2 - 36 = 0$

The equation $x^2 + 64 = 0$ can not be factored and solved. Explain what happens if you try to factor this equation.

Strength Builder

Fun with Crosswords

This puzzle can be completed individually or by groups of students.

Use the clues to fill in the puzzle. Many of the terms from the unit appear in the puzzle.

ACROSS

Number	Clue
1	arrangement of terms of a polynomial in order
3	property that allows a change in the order of terms
5	describes a square
7	type of polynomial with two perfect squares and only one subtraction sign
9	indicated by an exponent
10	has no addition or subtraction sign
11	alternative symbol for multiplication
12	has exactly two terms
14	opposite
16	helps in factoring some polynomials
17	can be used as a coefficient or an exponent
18	these can be added

DOWN

Number	Clue
1	result in addition
2	operations must follow this *or* opposite of chaos
3	numerical part of a monomial
4	combine terms
6	term with no variable
8	tells how many times to use a factor
9	may have one, two, or more terms
11	simplify common factors in numerator and denominator *or* to separate
13	based on the Associative Property, do this before adding or multiplying *or* a set
15	a mistake in calculation

The Puzzle

Unit 4

Review

Write each expression in standard form. Then write the most appropriate word in the box: *monomial, binomial, trinomial,* or *polynomial*.

1. $9 + 2x^2$

2. $x^3 + 3x + x^2$

3. $x^4 - 5x$

Add or subtract as indicated.

4. $(^-3y + 2) + (7y - 9) =$

5. $(5w^2 + 1) - (2w^2 + 6) =$

6. $(5x + xy) + (4xy + 9) =$

7. $(6x^3 + x^2 - 2x) - (x^2 - 4x + 4) =$

Multiply or divide as indicated.

8. $(x + 5)(x - 1) =$

9. $(7n + 4)^2 =$

10. $\frac{4x^2+8x}{2x} =$

11. $(z + 3)(z - 3) =$

12. $(2n^2)^3 =$

13. $4(a^2b)^2 =$

Factor each polynomial.

14. $x^2 - 10x + 25 =$

15. $x^2 - 100 =$

16. $x^2 + 18x + 81=$

Solve each equation by factoring.

17. $x^2 + 14x + 49 = 0$

18. $x^2 - 144 = 0$

19. $x^2 - 4x + 4 = 0$

Cumulative Review

Find the value of each expression.

1. $5 \times (^-2) =$

2. $^-9 \times (^-4) =$

3. $2^3 \times 1 =$

4. $^-8 \div (^-2) =$

5. $^-1 \div (1) =$

6. $^-12 + (^-4) =$

7. $|^-5| =$

8. $^-4 - 2 =$

9. $^-3 - (^-1) =$

Simplify. If the expression cannot be simplified, write *simplest form*.

10. $6x + 2x =$

11. $3ab + 16ab =$

12. $^-4x^2 + 10x^2 =$

13. $3a + (12a - 4a) =$

14. $^-9b + 3b^3 =$

15. $(a^2bc^3)^2 =$

16. $^-4(x^3y^2)^3 =$

17. $\frac{6a(b^3)^2}{4ab} =$
$a \neq 0, b \neq 0$

18. $\frac{x^4y^3z^6}{y^2z^4} =$
$y \neq 0, z \neq 0$

19. $x^{-5}(x^{10}y^3) =$

20. $3a^3(a^3b^2)^2 =$

21. $3ab + 9cd - 5ab =$

Write each number in scientific notation.

22. 63,000,000 =

23. 300,000 =

24. 701, 000, 000 =

Multiply or divide using scientific notation.

25. $210{,}000 \times 3{,}500 =$

26. $6{,}400{,}000 \div 32{,}000 =$

State the property that is used in each equation.

27. $x \cdot (y \cdot z) = (x \cdot y) \cdot z$

28. $^-5 + 5 = 0$

29. $4(x + y) = 4x + 4y$

Cumulative Review

Solve each equation.

30. $^-4n + 3 = 15$

31. $9t - 7 = 6t + 11$

32. $\frac{b}{3} + 7 = 12$

33. $6 - 3x = 10x - 7$

34. $^-4x + 5 = 13$

35. $7 + \frac{t}{2} = 0$

Solve each inequality and graph on a number line.

36. $x + 5 \leq 7$

$^-5\ ^-4\ ^-3\ ^-2\ ^-1\ 0\ 1\ 2\ 3\ 4\ 5$

37. $\frac{n}{-3} > {}^-1$

$^-5\ ^-4\ ^-3\ ^-2\ ^-1\ 0\ 1\ 2\ 3\ 4\ 5$

38. $2n + 6 \geq {}^-4$

$^-5\ ^-4\ ^-3\ ^-2\ ^-1\ 0\ 1\ 2\ 3\ 4\ 5$

39. Use the two points to find the slope of line AB. $A(^-3, 4)$ and $B(7, ^-5)$

40. Identify the slope and y-intercept of the line given by $y = \frac{2}{3}x - 7$.

41. Write the equation of a line with slope = 7 and y-intercept = $^-1$.

42. Evaluate the function $f(x) = {}^-4x + 3$ for $x = {}^-5$.

43. Determine whether $(^-2, 8)$ is a solution to $y = 5x + 3$.

44. Evaluate the function $f(x) = 2x - 6$ for $x = {}^-3$.

45. Solve the system by graphing.

$$\begin{cases} y = x + 2 \\ y = 3x - 4 \end{cases}$$

Solution:

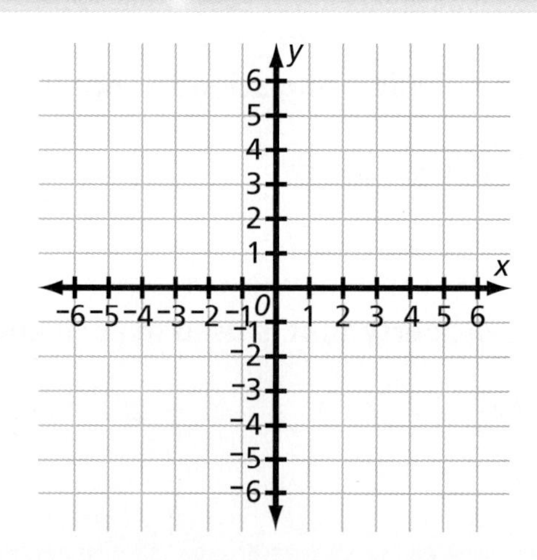

Write each polynomial in standard form.

46. $4x^3 - 2x + 3x^2 - 5$

47. $6 - 2x + 7x^2$

48. $4 - 6x$

Simplify.

49. $2(4t + 5) + (6t - 9) =$

50. $2(3a^2)^3 =$

51. $(2x^6 - 3x^2) + (5x^6 - 7x^2) =$

52. $3(6x - 2) + 2(3x + 1) =$

53. $(x + 3)(x - 2) =$

54. $(2x + 3)^2 =$

55. $\frac{8x^2 + 6x}{2x} =$

56. $(x - 4)(x - 1) =$

57. $^{-}3x^2(2x^4 + 4x - 2) =$

58. $(x - 4)^2 =$

Factor each polynomial.

59. $x^2 + 4x + 4 =$

60. $x^2 - 25 =$

61. $x^2 - 10x + 25 =$

62. $x^2 - 16 =$

63. $x^2 - 8x + 16 =$

64. $x^2 - 144 =$

Solve each equation by factoring.

65. $x^2 + 6x + 9 = 0$

66. $x^2 - 100 = 0$

67. $x^2 - 10x + 25 = 0$

68. $x^2 + 14x + 49 = 0$

69. $x^2 - 18x + 81 = 0$

70. $x^2 - 36 = 0$

Glossary

Absolute value (p. 6) An integer's distance from 0 on the number line; the symbol for absolute value is | |

$|3| = 3$ *and* $|^-3| = 3$

Additive inverses (p. 12) A number and its opposite; the sum is equal to zero

$^-8 + 8 = 0$

Associative Property of Addition (p. 36) States that the grouping of addends does not change the sum of those addends

$a + (b + c) = (a + b) + c$

Associative Property of Multiplication (p. 38) States that changing the grouping of factors does not change the product of those factors

$a \cdot (b \cdot c) = (a \cdot b) \cdot c$

Base (of an exponent) (p. 18) The number that is used as factor

$5^3 \leftarrow 5$ is the base

Binomial (p. 96) A polynomial with two terms

$6x + 9$

Boundary line (p. 82) A line used in graphing inequalities that separates the graph into different regions

Commutative Property of Addition (p. 36) States that changing the order of addends does not change the sum

$7 + 5 = 5 + 7$

Commutative Property of Multiplication (p. 38) States that changing the order of factors does not change the product of those factors

$5 \cdot 6 = 6 \cdot 5$

Constant of variation (p. 76) When two variables vary directly, the ratio of the variables equals a constant (k), which is the constant of variation.

Difference of two squares (p. 116) Pattern of a binomial having one perfect square subtracted from another perfect square

$x^2 - 4 = (x + 2)(x - 2)$

Distributive Property (p. 42) States that multiplying a sum by a number is the same as multiplying each addend by the number and then adding the products

$a(b + c) = ab + ac$

Domain (of a function) (p. 86) The set of all possible input values

Exponent (p. 18) The number that tells how many times the base is multiplied by itself

$5^3 \leftarrow 3$ is the exponent

$5^3 = 5 \cdot 5 \cdot 5$

Evaluate (p. 86) To find the value of an expression

Evaluate $f(x) = x + 2$ when $x = 4$.

$f(4) = 4 + 2 = 6$

Function (p. 86) A relation that has a single output value for each input value

Identity Property of Addition (p. 36) States that the addition of zero to a term does not change its value

$9 + 0 = 9$

Identity Property of Multiplication (p. 38) States that the product of any number and 1 is that number

$a \cdot 1 = a$, $2 \cdot 1 = 2$

Inequality (p. 58) Shows that two expressions are not equal and describes the relationship between the expressions

$5 < 15$, read *5 is less than 15*

Integers (p. 6) The set of whole numbers and their opposites

$\{\ldots {}^{-}3, {}^{-}2, {}^{-}1, 0, 1, 2, 3, \ldots\}$

Intercept (p. 74) The point where the graph of an equation crosses an axis

Isolate (a variable) (p. 44) To place the variable alone on one side of the equation

Inverse Property of Addition (p. 36) The sum of a term and its opposite is zero

${}^{-}x + x = 0$

Like terms (p. 16) Terms that have the same variable factors

$2x$ and $5x$ are like terms
$2x$ and $5xy$ are not like terms

Linear equation (p. 68) An equation whose solutions lie on a line in a graph

$y = 3x + 8$

Linear inequalities (p. 82) An inequality in which the variables are raised to the first power

$y < 3x + 8$

Monomial (p. 96) A number, variable, or the product of numbers and variables

5, $5x$, ${}^{-}4yz$

Multiplicative inverses (p. 38) Two numbers whose product is 1

$\frac{2}{3} \cdot \frac{3}{2} = 1$

Negative (p. 14) A number that is less than zero

Opposites (p. 12) Numbers that are the same distance from zero on the number line

The opposite of 4 is ${}^{-}4$.

Order of Operations (p. 40) A standard order for finding the value of a math expression: parentheses, exponents, multiplication and division, addition and subtraction

Origin (p. 68) The point where the x-axis and y-axis intersect

Perfect square trinomial (p. 114) A trinomial that results from squaring a binomial

$(x + 3)^2 = x^2 + 6x + 9$

Glossary

Polynomial (p. 96) The sum or difference of monomials

Positive (p. 14) A number that is greater than zero on the number line

Quadrants (p. 68) The four regions that are created when dividing the coordinate plane by the *x*- and *y*-axis

Range (of a function) (p. 86) The set of all possible output values of the function

Reciprocals (p. 28, p. 38) Two numbers whose product is 1

$\frac{2}{3} \cdot \frac{3}{2} = 1$

Rise (p. 72) The vertical change from one point to another on a coordinate plane

Run (p. 72) The horizontal change from one point to another on a coordinate plane

Scientific Notation (p. 20) A way to write large numbers that shows how many times a smaller number is multiplied by 10

$50{,}000 = 5.0 \times 10^4$

Simplest form (p. 16) An expression in which no like terms exist

$6x + 3y - 7$

Slope (p. 72) The ratio of the vertical change to the horizontal change

Slope-intercept form (p. 74) A form of a linear equation, $y = mx + b$, where m is the slope and b is the y-intercept

Solution set (p. 78) All the possible values of the variables that make an equation true

Standard form (p. 96) Terms that are arranged from left to right with exponents in order from greatest to least

$x^2 + 3x - 8$

System of equations (p. 84) Two or more equations that use two or more variables

Trinomial (p. 96) The sum or difference of three monomials

Vary directly (p. 76) The value of one variable increases when the value of another variable increases

***x*-axis** (p. 68) The horizontal axis of the coordinate plane

***y*-axis** (p. 68) The vertical axis of the coordinate plane

Zero Property of Multiplication (p. 28) States that when zero is a factor, the product is zero

$5 \times 0 = 0$